UPCO's
Review of
CHEMISTRY

REVISED

Robert M. Capie
Former Science Department Head
Port Jefferson Public Schools
Port Jefferson, New York

Revisions by
Ross MacKinnon
Chemistry Teacher,
Scotia Glenville High School
Scotia, New York

United Publishing Company, Inc.
76 Exchange Street
Albany, N. Y. 12205

Note:
A black line in the margin indicates optional areas that are not part of the core chemistry curriculum. These areas are tested only in Part II of the Regents examination.

ISBN 0-937323-04-7

1 2 3 4 5 6 7 8 9 0

CONTENTS

UNIT 1. MATTER AND ENERGY

DEFINITION OF CHEMISTRY

Chemistry is the study of the composition, structure, and properties of matter, the changes which matter undergoes, and the energy accompanying those changes.

ENERGY

Although there are many more sophisticated definitions, **energy** is probably best defined as the capacity to do work. **Work** is done whenever a force is used to change the position and/or motion of matter in a system.

All energy can be classified into two types—potential energy and kinetic energy. **Potential energy** is related to change in position. **Kinetic energy** is related to change in motion.

When we roll a ball up a hill, we expend kinetic energy. The ball, however, acquires potential energy as a result of the work done on it. At the top of the hill, the ball at rest has potential energy due to its position. If the ball rolls down the hill, the potential energy will change to kinetic energy of motion.

Forms of Energy

Energy, whether potential or kinetic, is associated with physical phenomena. Traditionally, different *forms* of energy are identified as follows:

- **Thermal energy** is the total amount of internal energy an object has.
- **Chemical energy** is the energy associated with chemical change.
- **Light energy** is the energy associated with electromagnetic radiation.
- **Nuclear energy** is the energy associated with changes in the masses of atoms.

Conservation of Energy

During ordinary chemical or physical changes, energy is conserved. Energy may be changed from one form to another, or transferred from one body or system to another, but the total amount of energy remains constant.

Any activity that releases energy is said to be **exothermic.** Activities that require energy are said to be **endothermic**.

Measurement of Energy

Temperature is a measure of the *average* kinetic energy of the particles in a substance. Two things should be emphasized in this definition:

1. The word *average* indicates that temperature might be calculated by adding the kinetic energies of all the particles and dividing by the number of particles. Under most conditions, each particle possesses its own kinetic energy, which may be quite different from that of any of its neighboring particles.

2. The word *heat* does *not* appear in the definition, although the temperature may be affected by the amount of heat in the system. A rising temperature is associated with the addition of heat to the system; falling temperature is associated with a loss of heat. Whenever two bodies with different temperatures get close enough to each other, *heat* flows from the body with the higher temperature to the one with the lower temperature until the two bodies are at the same temperature.

Thermometry. Temperatures are usually measured by thermometers. Most thermometers consist of a liquid confined inside a glass tube. A change in temperature produces a resultant change in the volume of the liquid. This change is seen as a change in the height of the column of liquid in the tube. For practical purposes, a substance that remains a liquid over a broad temperature range is used. The most commonly used liquids are mercury and alcohol.

Thermometers are calibrated using two special temperatures of water measured at normal, or standard, atmospheric pressure. One temperature is measured while the thermometer is in water that is being cooled. The temperature at which the mercury stops falling and the water begins to change to ice is the **freezing point** of water. This point is also known as the solid-liquid equilibrium temperature, and is zero degrees Celsius (0°C).

The other temperature used to calibrate a thermometer is measured when the thermometer is placed in water that is being heated. The temperature at which the mercury stops rising and the water starts to turn to a vapor is the **boiling point,** or liquid-gas equilibrium temperature of water. This temperature is one hundred degrees Celsius (100°C).

The distance between the freezing point temperature (0°C) and the boiling point temperature (100°C) is divided into 100 equal intervals called Celsius degrees (°C). An "absolute" temperature scale was proposed by Kelvin. Called the Kelvin scale, this scale is based on ascribing an absolute zero temperature value for no kinetic energy. At this temperature, all particle motion would stop. This absolute zero is equivalent to 273 degrees below zero on the Celsius scale (–273°C). The size of each Kelvin degree, referred to as a Kelvin, is the same as that of each Celsius degree. Thus,

$$K = °C + 273$$

The freezing point of water is 273K and the boiling point of water is 373K at standard atmospheric pressure. Note that no degree symbol (°) is used when writing Kelvin temperatures.

Calorimetry. A **calorimeter** is a device used to measure heat of reaction. The calorimeter uses the change in temperature of water to measure the amount of heat produced during exothermic processes and the amount of heat absorbed during endothermic processes. The process is allowed to proceed in a container that is in contact with a known mass of water. The temperature of the water is measured before and after the reaction (process), and the change in temperature is determined.

Temperature change must be converted to units of heat energy. One basic unit of heat energy is the **calorie**. One calorie is defined as the quantity of heat needed to raise the temperature of 1 gram of water by 1 Celsius degree. Thus, the energy involved in changing the temperature of a known mass of water can be found by multiplying the number of grams of water (mass) times the change in the number of Celsius degrees (temperature change). Temperature change is shown by the notation ΔT. Heat exchange in the calorimeter can be expressed mathematically as

$$\text{calories} = m \times \Delta t$$

Substances other than water can be used in a calorimeter provided that we account for the differences in the amount of heat required to change the temperature of the new substance by 1°C. This amount of heat is known as the **specific heat** of the substance. Specific heat is represented by the letter c. Thus, a general formula used in calorimetry is: Heat gained or lost = mass of liquid × specific heat of the liquid × temperature change of the liquid. Symbolically, this can be written.

$$\Delta H = m \times c \times \Delta t$$

SAMPLE PROBLEMS

How many calories are absorbed by 30.0g of water when its temperature is raised from 20.°C to 40.°C?

Solution:

Given:	mass (water) = 30.0g
	$\Delta t = 40.°C - 20.°C = 20.°C$
Known:	$c = 1$ cal/gC° (for water)
Find:	heat gained
Equation:	Heat gained = $mc\Delta t$
	= 30.0g × 1 cal/g°C × 20.°C
	= 6.0×10^2 calories

A reaction chamber inside a calorimeter contains 150.g of water at 19°C. A reaction is permitted to take place in the chamber. After the reaction, the temperature of the water is 29°C. How many calories were released by the reaction?

Heat released = mass × specific heat × temperature change
   = $mc\Delta t$
   = 150.g × 1 cal/g°C × 10.°C
   = 1.5×10^3 calories

MATTER

The traditional definition of **matter** is anything that has mass and volume. The term matter is ordinarily used when referring to any material found in nature. These materials are found in what seems to be an infinite variety of forms, and are classified in many ways. Two groups into which all matter may be placed are homogeneous matter and heterogeneous matter.

- **Homogeneous matter** includes all those materials having uniform characteristics throughout a given sample. This group includes elements, compounds, and solutions.

- **Heterogeneous matter** is not uniform in its composition. Its parts are not alike.

Scientists also classify matter into the two groups—**pure substances** and **mixtures** of substances. Pure Substances are almost always homogeneous. Mixtures may be homogeneous or heterogeneous. Solutions are examples of homogeneous mixtures.

Elements, Compounds and Mixtures

Elements are the simplest kinds of substances. They are made up of atoms that are essentially all alike. They cannot be broken down into simpler substances by chemical change. Carbon, hydrogen, oxygen, iron, and sulfur are examples of elements.

Compounds are substances made up of two or more elements that are chemically combined. These substances *can* be broken into their component elements by chemical change. The properties of compounds are generally quite different from those of the elements of which they are composed. H_2O, CO_2 and $NaCl$ are examples of compounds.

Mixtures are blends of substances whose components might be elements, compounds, or both. Mixtures can usually be separated by simple physical means, and their properties are ordinarily a combination of the properties of their components. Most mixtures are heterogeneous. Solutions are examples of homogeneous mixtures.

QUESTIONS

1. Which of the following substances can *not* be decomposed by chemical change? (1) sulfuric acid (2) ammonia (3) water (4) argon
2. An example of a heterogeneous mixture is (1) soil (2) sugar (3) carbon monoxide (4) carbon dioxide

3. Which substance is composed of atoms that all have the same atomic number? (1) magnesium (2) methane (3) ethane (4) ethene

4. At 1 atmosphere of pressure, the fixed temperature points on a Celsius thermometer are located on the basis of (1) the ice/water equilibrium temperature, only (2) the water/steam equilibrium temperature, only (3) both the ice/water and the water/steam equilibrium temperatures (4) neither the ice/water nor the water/steam equilibrium temperatures

5. The boiling point of water at standard pressure is (1) 0.000 K (2) 100. K (3) 273 K (4) 373 K

6. Which temperature represents absolute zero? (1) 0 K (2) 0°C (3) 273 K (4) 273°C

7. The combustion of propane is best described as an (1) endothermic chemical change (2) endothermic physical change (3) exothermic chemical change (4) exothermic physical change

8. The reaction of hydrogen and oxygen to form water is best described as (1) exothermic, because energy is released (2) endothermic, because energy is released (3) exothermic, because energy is absorbed (4) endothermic, because energy is absorbed

9. The average kinetic energy of the molecules of an ideal gas is directly proportional to the (1) number of moles present (2) volume occupied by individual gas molecules (3) temperature measured on the Kelvin scale (4) pressure at standard temperature

10. When a sample of a gas is heated at constant pressure, the average kinetic energy of its molecules (1) decreases, and the volume of the gas increases (2) decreases, and the volume of the gas decreases (3) increases, and the volume of the gas increases (4) increases, and the volume of the gas decreases

11. In which beaker would the particles have the highest average kinetic energy?

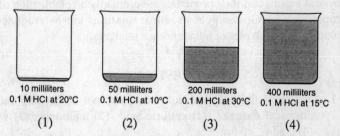

10 milliliters	50 milliliters	200 milliliters	400 milliliters
0.1 M HCl at 20°C	0.1 M HCl at 10°C	0.1 M HCl at 30°C	0.1 M HCl at 15°C
(1)	(2)	(3)	(4)

12. Which sample could represent a homogeneous mixture? (1) $CO_2(aq)$ (2) $CO_2(s)$ (3) $CO_2(l)$ (4) $CO_2(g)$

13. The temperature of a substance is a measure of its particles'
     (1) average potential energy   (2) average kinetic energy
     (3) enthalpy   (4) entropy
14. How many kilocalories are equivalent to 1.0 calorie?
     (1) 0.001 kcal   (2) 0.01 kcal   (3) 1000 kcal   (4) 10,000 kcal
15. What is the total number of calories of heat energy absorbed when
     the temperature of 200 grams of water is raised from 10°C to 40°C?
     (1) 30 cal   (2) 200 cal   (3) 6000 cal   (4) 8000 cal
16. How many grams of water will absorb a total of 600. calories of
     energy when the temperature of the water changes from 10.0°C to
     30.0°C?   (1) 10.0 g (2) 20.0 g   (3) 30.0 g   (4) 60.0 g
17. How many calories of heat energy are absorbed in raising the
     temperature of 10. grams of water from 5.0°C to 20.°C?
     (1) 2.5×10^2 (2) 2.0×10^2 (3) 1.5×10^2 (4) 5.0×10^1
18. Which energy change occurs during the burning of magnesium
     ribbon?
     (1) chemical energy → light energy
     (2) chemical energy → electrical energy
     (3) electrical energy → chemical energy
     (4) electrical energy → light energy
19. Given the equation: $I + I \rightarrow I_2 + 35$ kcal
     This equation shows that the formation of an iodine molecule is an
     (1) exothermic process in which energy is absorbed
     (2) exothermic process in which energy is released
     (3) endothermic process in which energy is absorbed
     (4) endothermic process in which energy is released
20. Which change of phase is exothermic?   (1) $H_2O(s) \rightarrow H_2O(g)$
     (2) $CO_2(s) \rightarrow CO_2(l)$ (3) $H_2S(g) \rightarrow H_2S(l)$ (4) $NH_3(l) \rightarrow NH_3(g)$
21. Which formula represents a binary compound?   (1) NH_4NO_3
     (2) CH_4 (3) CH_3COCH_3 (4) $CaCO_3$

PHASES OF MATTER

Matter exists in nature in three **phases**—solid, liquid, and gas. These
terms might be used to describe materials as they are found in nature
under ordinary conditions. Water is most often found in liquid form, air is
a mixture of gases, and the rocky portion of the earth consists mostly of
materials in solid form.

In this consideration of phases of matter, we shall be concerned with
the transfer of heat as different materials undergo *phase changes*—that
is, change from one phase to another. Since water is familar to us in all
three phases, we will use it for our model.

Start with a sample of ice at some temperature lower than 0°C. If we were to supply heat to this sample at a constant rate, the temperature of the ice would rise uniformly until it began to melt, or change phase, at the melting point (0°C). Even though we continued to add heat at a constant rate, the temperature of the resulting ice-water mixture would not increase. It would remain at 0°C until all of the ice had changed to liquid water.

Once all the water was in the liquid phase, its temperature would rise uniformly until it began to boil at 100°C. Once again, the temperature of the liquid-vapor mixture would remain at 100°C until all the liquid had changed to a gas. At that point, the gas would undergo a uniform increase in temperature.

Gases

Gases are transparent, they can be compressed, and they expand without limit. Gases assume the shape and volume of their container. The volume of a given mass of a gas is the volume of the container that holds it.

Gases exert pressure on the walls of a container. Pressure is most often expressed in units of force per unit area of surface.

The earth's atmosphere exerts pressure, called air pressure, or atmospheric pressure. Air pressure is measured with a barometer. One type of barometer consists of a glass tube about one meter long, sealed at one end and partly filled with mercury. The open end of the tube is placed in a reservoir of mercury that is exposed to the atmosphere. The weight of the air exerts a pressure on the surface of the mercury in the reservoir. This pressure supports a column of mercury in the glass tube.

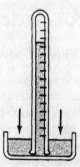

Figure 1-1. Mercury Barometer. The greater the air pressure (arrows) the higher the mercury rises in the tube.

The height of the mercury column is directly related to the amount of pressure exerted by the atmosphere on the mercury in the reservoir. At 0°C at sea level, the atmospheric pressure supports a column of mercury 760 mm high. In honor of Torricelli, who invented the mercury barometer, this pressure is often called 760 torr. This value—760 torr or 760 mm of mercury—is also known as one atmosphere of pressure, or standard pressure.

Boyle's Law. As long as the temperature of a given mass of a gas is unchanged (is constant), an increase in pressure on the gas causes a decrease in its volume; a decrease in pressure causes an increase in its volume. This kind of relationship is referred to as an inverse proportion. **Boyle's law** states that, at constant temperature, the volume of a given mass of gas varies inversely with the pressure exerted on it. This law can be expressed mathematically as

$$\frac{V_1}{V_2} = \frac{P_2}{P_1}$$

where V_1 and P_1 represent original volume and pressure and V_2 and P_2 represent the new volume and pressure.

Boyle's law can also be written as:

$$V_1 P_1 = V_2 P_2 \text{ OR } V_2 = V_1 \times \frac{P_1}{P_2}$$

From this mathematical expression, we can see that the product of the volume and pressure of a given mass of a gas at constant temperature is always the same. Thus, we can write

$$PV = k$$

where k is a constant value.

SAMPLE PROBLEM

The volume of a gas is 342 mL at 25.0° C and 730. torr. Find the volume of the gas at 146 torr, temperature remaining unchanged.

Solution:

1. Set up a table.

	1	2
V	342 mL	x
P	730. torr	146 torr

2. Write equation.

$$V_2 = V_1 \times \frac{P_1}{P_2}$$

3. Substitute.

$$V_2 = 342 \text{ mL} \times \frac{730. \text{ torr}}{146 \text{ torr}}$$

$$= 342 \text{ mL} \times \frac{730.}{146}$$

$$= \quad \textbf{1710 mL}$$
[Remember mL = cm³]

Charles' law. Charles' law explains the relationship between the volume of a given mass of a gas and its absolute (Kelvin) temperature. This law states that at constant pressure, the volume of a given mass of a gas varies *directly* with its Kelvin temperature. In other words, at constant pressure, as temperature increases, volume increases; as temperature decreases, volume decreases.

Charles' law can be represented as

$$\frac{V_1}{V_2} = \frac{T_1}{T_2} \quad \text{OR} \quad \frac{V_1}{T_1} = \frac{V_2}{T_2} \quad \text{OR} \quad V_2 = V_1 \times \frac{T_2}{T_1}$$

This relationship between volume and temperature predicts that at a temperature of 0 K (absolute zero), the volume of a gas will be zero. This prediction is not realistic, because gas molecules have both mass and volume. These properties will not change, regardless of temperature. Thus, the volume of a gas could not reach zero.

In reality, the volume of the *space between* the molecules of gas approaches zero. All gases will liquefy and most will solidify before reaching 0 K at 1 atmosphere. Additionally, absolute zero is a theoretical value. It cannot be attained in any real system, although temperatures nearly that cold have been reached.

SAMPLE PROBLEM

A gas occupies a volume of 560. cm^3 at a temperature of 100.°C. To what temperature must the gas be lowered if it is to occupy 400. cm^3, pressure remaining unchanged?

Solution:

Change temperature to Kelvin

1.   100. + 273 = 373 K

2.

	1	2
V	560. cm^3	400. cm^3
T	373 K	x

Remember: Always use Kelvin in Charles' Law

3. $\dfrac{V_1}{V_2} = \dfrac{T_1}{T_2}$ OR $T_2 = \dfrac{T_1 V_2}{V_1}$

4. $T_2 = \dfrac{(373K)(400. cm^3)}{560. cm^3}$

$= \dfrac{(373K)(400.)}{560.}$

$= \mathbf{266 \ K \ or \ -7°C}$

Combined Gas Laws. In most systems, both temperature and pressure vary. In such cases, Boyle's law and Charles' law must be applied together. This **combined gas law** can be represented as

$$\frac{V_1 P_1}{T_1} = \frac{V_2 P_2}{T_2} \quad \text{OR} \quad V_1 P_1 T_2 = V_2 P_2 T_1$$

Thus, the product of the original volume and pressure and the new Kelvin temperature is equal to the product of the new volume and pressure and the original Kelvin temperature.

The symbols in the mathematical expression for the combined gas law can be rearranged so as to isolate a single factor on one side of the equation. For example, the expression isolating the new volume (V_2) would be written

$$V_2 = V_1 \times \frac{P_1}{P_2} \times \frac{T_2}{T_1}$$

Similarly, the expression isolating the original pressure would be written

$$P_1 = P_2 \times \frac{V_2}{V_1} \times \frac{T_1}{T_2}$$

It is suggested that the equation for the combined gas law be used for solving all problems involving gases. The equation is:

$$\frac{V_1 P_1}{T_1} = \frac{V_2 P_2}{T_2}$$

The advantages of using this equation are:

- Only one equation is needed.
- All initial values are on the left of the equation; all new values are on the right of the equation.
- In cases where either temperature or pressure is unchanged, it can be ignored (or included); its value will not affect the problem.

It is also recommended that tables be set up and used. The following model can be used:

	1 Initial conditions	2 New or final conditions
Volume		
Pressure		
Temperature K		

SAMPLE PROBLEM

A sample of gas had a volume of 200. L under a pressure of 300. torr and a temperature of 20.° C. What volume would the sample occupy at a pressure of 250. torr and a temperature of 30.°C?

Solution:

1. Convert temperatures to Kelvin:

    20.°C + 273 = 293K; 30.°C + 273 = 303K

2. Set up a table.

	1	2
V	200. L	x
P	300. torr	250. torr
T	293 K	303K

3. Write equation:

$$\frac{V_1 P_1}{T_1} = \frac{V_2 P_2}{T_2} \quad \text{or} \quad V_2 = V_1 \times \frac{P_1}{P_2} \times \frac{T_2}{T_1}$$

4. Substitute:

$$V_2 = 200.\,L \times \frac{300.\,torr}{250.\,torr} \times \frac{303K}{293K}$$

$$= 248\ L$$

Standard Temperature and Pressure. Gas volumes are influenced considerably by changes in temperature and pressure. Thus, when working with gases, it is convenient to define standard reference conditions. By convention, 0°C (273K) and 760 torr (1 atm) represent **standard temperature** and **pressure**, often designated as **STP**. When dealing with a gas volume, STP is assumed unless otherwise indicated.

Dalton's Law of Partial Pressures. The pressure exerted by each of the gases in a mixture of gases is called the *partial pressure* of that gas. The total pressure exerted by the mixture of gases is the sum of the partial pressures of the gases that make up the mixture. This relationship, known as **Dalton's law of partial pressures,** is generally expressed as

$$P_{total} = P_1 + P_2 + P_3 \ldots$$

where P_1, P_2, and P_3 represent the partial pressures exerted by the gases in the mixture.

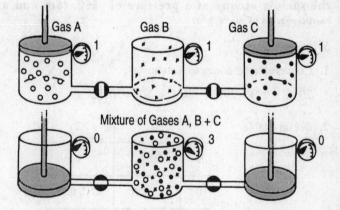

Figure 1-2. Dalton's Law of Partial Pressures. The three containers (top) each holds an equal volume of a different gas at a pressure of 1 atmosphere. (Bottom) The gases from the outer containers have been added to the gas in the center. The pressure is 1 atm. + 1 atm. + 1 atm. or 3 atmospheres.

Graham's Law. Gases expand, or spread out, to occupy the volume available to them. This "spreading out" of a gas is called **diffusion.** The rate at which a gas diffuses is inversely related to the square root of the mass of its molecules. In other words, the greater the mass, the slower the speed.

When comparing two gases at the same temperature and pressure, the gases will diffuse at a rate inversely proportional to the square roots of their molecular masses (or densities). This relationship, known as Graham's law, can be represented as

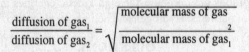

$$\frac{\text{diffusion of gas}_1}{\text{diffusion of gas}_2} = \sqrt{\frac{\text{molecular mass of gas}_2}{\text{molecular mass of gas}_1}}$$

Avogadro's Hypothesis. Two equal volumes of gases at the same temperature and pressure must have equal numbers of particles, even though their masses are not the same. For example, at the same temperature and pressure, one liter of nitrogen gas contains the same number of particles as one liter of hydrogen gas, even though the mass of the nitrogen gas is fourteen times that of the hydrogen gas.

Moles and Molar Volume. Atoms, molecules, electrons, ions, and other particles that chemists deal with are extremely small. When comparing quantities of different substances, chemists use a unit that contains an extremely large number of particles. The unit, called a mole, contains 6.02×10^{23} particles. This number is known as Avogadro's number. At STP, one mole of any gas occupies a volume of 22.4 liters, which is known as a molar volume.

Kinetic Theory of Gases. Since so many gases exhibit the same behavior, a model was developed to help explain this similarity. This kinetic theory of gases is based on these assumptions.

1. All gases are composed of tiny, individual particles that are in continuous motion. These particles move rapidly, randomly and in straight lines.
2. When particles collide with one another, energy is transferred, *without loss,* from one particle to the other. Therefore, the net total energy of the system remains constant.
3. Compared to the distances between them, the particles are so small that their volumes are considered to be zero.
4. The particles have no attraction for one another.

The gases described in this model are considered to be "ideal" gases. In reality, no gas meets all of the qualities of this model.

Deviations from the Gas Laws. Real gases do not behave exactly as predicted by the "ideal" gas of the model. Deviations from the ideal behavior are due to the fact that gas particles *do have volume* and they *do exert some attraction* for one another. Deviations from the gas laws are least obvious among light gases at high temperatures and low pressures. These conditions are optimum for high kinetic energies and maximum separation of particles. Hydrogen and helium are closest to "ideal" gases.

QUESTIONS

1. Which substance takes the shape of and fills the volume of any container into which it is placed.   (1) $H_2O(l)$ (2) $I_2(s)$ (3) $CO_2(g)$ (4) $Hg(l)$

2. If the pressure on 36.0 milliliters of a gas at STP is changed to 190. millimeters of mercury at constant temperature, the new volume of the gas is   (1) 9.00 mL   (2) 126 mL   (3) 144 mL   (4) 226 mL

3. At constant temperature the pressure on 8.0 liters of a gas is increased from 1 atmosphere to 4 atmospheres. What will be the new volume of the gas? (1) $1.0\,L$ (2) $2.0\,L$ (3) $32\,L$ (4) $4.0\,L$

4. Which sample of methane gas contains the greatest number of molecules at standard temperature? (1) 22.4 liters at 1 atmosphere (2) 22.4 liters at 2 atmospheres (3) 11.2 liters at 1 atmosphere (4) 11.2 liters at 2 atmospheres

5. The pressure on 200. milliliters of a gas is decreased at constant temperature from 900. torr to 800. torr. The new volume of the gas, in milliliters, is equal to

(1) $200. \times \dfrac{900.}{800.}$ $\qquad$ (2) $200. \times \dfrac{800.}{900.}$

(3) $800. \times \dfrac{200.}{900.}$ $\qquad$ (4) $800. \times \dfrac{900.}{200.}$

6. The diagram below represents a gas confined in a cylinder fitted with a movable piston.

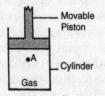

As the piston moves toward point A at a constant temperature, which relationship involving pressure (P) and volume (V) is correct?
(1) $P + V = k$ (2) $P - V = k$ (3) $P \div V = k$ (4) $P \times V = k$

7. What is the difference in pressure between a gas at 720 mm Hg and a gas at standard pressure? (1) 0 mm Hg (2) 40 mm Hg (3) 273 mm Hg (4) 760 mm Hg

8. A 300.-milliliter container that is filled with 100. milliliters of oxygen and 200. milliliters of hydrogen has a total pressure of 750. millimeters of mercury. What is the partial pressure of the oxygen? (1) 100. mmHg (2) 200. mmHg (3) 250. mmHg (4) 500. mmHg

9. Gases in the amounts shown in the table below were added to an empty 22.4-liter container at 0°C.

GAS	QUANTITY (moles)
O_2	0.500
N_2	0.250
He	0.125

How much argon gas should be added to the container at 0°C to produce a total pressure of 1.00 atmosphere? (1) 1.12 moles (2) 2.24 moles (3) 0.375 mole (4) 0.125 mole

10. Which sample of water has the greatest vapor pressure? (1) 100 mL at 20°C (2) 200 mL at 25°C (3) 20 mL at 30°C (4) 40 mL at 35° C

11. A 0.500-mole sample of a gas has a volume of 11.2 liters at 273 K. What is the pressure of the gas? (1) 11.2 torr (2) 273 torr (3) 380. torr (4) 760. torr

12. Equal volumes of $SO_2(g)$ and $O_2(g)$ at STP contain the same number of (1) atoms (2) molecules (3) electrons (4) protons

13. At STP, 1 liter of $H_2(g)$ and 1 liter of He(g) have the same (1) mass (2) density (3) number of atoms (4) number of molecules

14. Under the same conditions of temperature and pressure, which of the following noble gases diffuses most rapidly? (1) Ne (2) Ar (3) Kr (4) Xe

15. Which gas would diffuse most rapidly under the same conditions of temperature and pressure? (1) gas A, molecular mass = 4 (2) gas B, molecular mass = 16 (3) gas C, molecular mass = 36 (4) gas D, molecular mass = 49

16. One reason that a real gas deviates from an ideal gas is that the molecules of the real gas have (1) a straight-line motion (2) no net loss of energy on collision (3) a negligible volume (4) forces of attraction for each other

17. Which phase change is endothermic? (1) liquid to gas (2) liquid to solid (3) gas to liquid (4) gas to solid

18. A flask contains a mixture of $N_2(g)$ and $O_2(g)$ at STP. If the partial pressure exerted by the $N_2(g)$ is 480. millimeters of mercury (mm Hg), the partial pressure of the $O_2(g)$ is (1) 280. mm Hg (2) 373 mm Hg (3) 480. mm Hg (4) 760. mm Hg

19. Which of the following gases would have the *slowest* rate of diffusion when all of the gases are held at the same temperature and pressure? (1) N_2 (2) NO (3) O_2 (4) CO_2

20. A 100.-milliliter sample of helium gas is placed in a sealed container of fixed volume. As the temperature of the confined gas increases from 10.°C to 30.°C, the internal pressure (1) decreases (2) increases (3) remains the same

Liquids

Liquids have definite volumes but do not have definite shapes. Like gases, liquids take on the shape of their container.

Evaporation. The force of attraction among particles of a liquid is much greater than that of a gas. The particles in a liquid are in constant motion, the speed depending on temperature. Particles that have a greater-than-average speed can overcome forces of attraction and leave the surface of the liquid to enter the gas, or vapor, phase. The term **vapor** is used to refer to the gas phase of a substance that is usually a solid or liquid at room temperature.

The escape of particles from the surface of a liquid to form the vapor is called **evaporation.** In an open container, evaporation continues until all the liquid has vaporized.

Condensation. If a liquid is placed in a closed container at a given temperature, only a small quantity will evaporate. It is impossible for particles that escape into the gas phase to leave the container. As more and more particles enter the gas phase, the likelihood that some of them will strike the liquid surface and return to the liquid phase increases. The change from vapor to liquid is called **condensation.** In a closed system, the rate of condensation eventually becomes equal to the rate of evaporation at the given temperature.

Vapor Pressure. In any closed system containing a liquid, the vapor produced by evaporation exerts a pressure, which is called the **vapor pressure** of the liquid. Every liquid has its own characteristic vapor pressure. Vapor pressure increases as temperature increases.

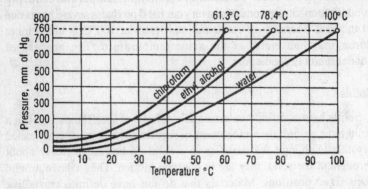

Figure 1-3. Vapor Pressure Curves for Three Substances

Boiling Point. Upon being heated, most liquids eventually boil. Boiling is characterized by the formation of bubbles of vapor in the body of the liquid, as the change from liquid to vapor takes place within the body of the liquid, as well as at the surface.

As described earlier, vapor pressure increases with an increase in temperature. The temperature at which the vapor pressure of a liquid is equal to the pressure pushing down on the surface of the liquid is called the **boiling point** of the liquid. The **normal boiling point** of a liquid is defined as the temperature at which the vapor pressure of the liquid is one atmosphere, or 760 torr.

In most circumstances, when reference is made to the boiling point of a substance, it is the normal boiling point that is indicated. However, the temperature at which a liquid will boil can be changed by changing the pressure pushing down on the surface of the liquid. For example, the vapor pressure of water at 80°C is 355 torr. Thus, if the pressure on the surface of a sample of water is reduced to 355 torr, the water will boil at 80°C. Conversely, increasing the pressure raises the boiling point temperature.

Heat of Vaporization. The particles that escape from the surface of a liquid are those having the highest kinetic energy. The remaining particles have lower average kinetic energy. Thus, heat must be added to the liquid to maintain constant temperature. Vaporization, then, is an endothermic process. The **heat of vaporization** of a liquid is the number of calories per gram of liquid that must be added in order to maintain a constant temperature while vaporization, or boiling, occurs. For water at its normal boiling point of 100°C, the heat of vaporization is approximately 540 calories per gram.

Heat of vaporization is also the number of calories per gram that must be *removed* in order to maintain constant temperature during condensation. Steam condensing on your hand produces severe burns due to the heat liberated on condensation. Steam at 100°C is much more damaging than water at the same temperature. The process of condensation is exothermic.

Solids

Solids have a definite volume and a definite shape. The particles in a solid have an orderly arrangement in regular geometric patterns, called crystals. Although the particles in a solid do not move freely about throughout the solid, they are in constant motion. They vibrate around their fixed positions. Materials that do not have definite crystalline structures, such as glass, paraffin, and plastic, are called amorphous solids. They are not considered to be true solids, but are supercooled liquids.

Freezing-Melting Point. When a liquid is cooled (heat is removed), a temperature is eventually reached at which the liquid begins to freeze. It changes to a solid. This temperature, which remains constant until all the liquid has solidified at 1 atmosphere pressure, is called the **freezing point** of the liquid. While the liquid is cooling, the average kinetic energy of its particles decreases until it is low enough for the attractive forces to be able to hold the particles in the fixed positions characteristic of the solid phase.

Alternately, warming a solid eventually causes the solid to melt. It begins to change to a liquid. During this phase change, the temperature remains constant until all the solid has liquefied at 1 atmosphere pressure. This temperature is called the **melting point** of the solid. For any given substance, the melting point temperature is exactly the same as the freezing point temperature. The only difference is in the direction of approach.

The melting-freezing point of a substance may also be defined as the temperature at which the liquid phase and the solid phase exist in equilibrium.

Heat of Fusion. Melting is an endothermic process. In order to maintain a constant temperature during this phase change, heat must be continually added to the system. The amount of heat needed to change a unit mass of a substance from solid to liquid at constant temperature and 1 atmosphere pressure is called the **heat of fusion** of that substance. The heat of fusion of ice at 0°C and 1 atmosphere is close to 80 calories per gram. To change one gram of ice at 0°C and 1 atmosphere pressure to one gram of water at the same conditions, approximately 80 calories of heat must be added. Conversely, to change 1 gram of liquid water at 0°C and 1 atm to 1 gram of ice at the same conditions, approximately 80 calories of heat must be removed.

A graph of temperature vs time will make it easier to visualize the phase change process. Figure 1-4 is such a graph for a substance with a melting point of 50°C and a boiling point of 110°C. The portion of the curve between A and B represents the time the solid phase is being heated. Note the uniform increase in temperature during this interval. The portion of the curve between B and C represents the time the substance is undergoing a phase change, from solid to liquid (heat of fusion). Note that there is no change in temperature during this interval, even though heat is being added at a constant rate. The heat is being absorbed in the process of the phase change.

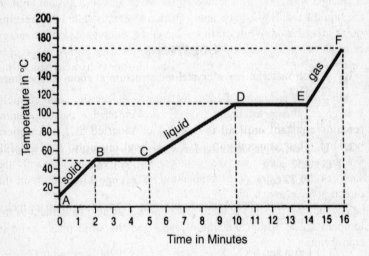

Figure 1-4. Heating Curve

That part of the curve between C and D represents the time interval that the liquid is being heated. Note the uniform increase in temperature during this interval. At point D, the temperature stops rising.

The interval between D and E represents the time interval in which the liquid is changing to a gas (heat of vaporization). Once again, there is no change in temperature during this time. At point E, all of the liquid has changed to a gas, and the temperature of the gas begins to rise at a uniform rate.

Such a graph is called a *heating curve*. If you were to read this graph from right to left, it would be a *cooling curve*.

Sublimation. Under certain conditions, it is possible for a substance to change from a solid directly into the gas phase without obviously passing through the liquid phase. This solid-to-gas change is called **sublimation.** Iodine and carbon dioxide are examples of substances that may undergo sublimation.

QUESTIONS

1. Which material has a crystalline structure at room temperature (20°C)?
   (1) water   (2) glass
   (3) nitrogen   (4) sucrose

2. The heat of vaporization for water at its normal boiling point is
   (1) 0.52 cal/g   (2) 1.86 cal/g
   (3) 79.72 cal/g   (4) 539.4 cal/g

3. Which conditions of pressure and temperature exist when ice melts at its normal melting point?

   (1) 1 atm and 0°C   (2) 760 atm and 0°C
   (3) 1 atm and 0 K   (4) 760 atm and 273 K

4. The graph below represents the uniform cooling of water at 1 atmosphere, starting with water as a gas above its boiling point.

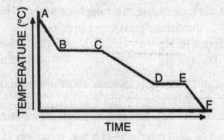

Which segments of the cooling curve represent the fixed points on a thermometer?
(1) *AB* and *CD*   (2) *BC* and *DE*
(3) *AB* and *EF*   (4) *CD* and *EF*

5. The heat of fusion of a substance is the energy measured during a
(1) phase change   (2) temperature change
(3) chemical change   (4) pressure change

6. Which material exists as a supercooled liquid at STP?
(1) salt   (2) sand
(3) diamond   (4) glass

7. Which substance will sublime at room temperature (20°C)?
(1) $C_{12}H_{22}O_{11}(s)$ (2) $C_6H_{12}O_6(s)$
(3) $SiO_2(s)$ (4) $CO_2(s)$

8. The temperature of 50 grams of water was raised to 50°C by the addition of 1,000 calories of heat energy. What was the initial temperature of the water?
(1) 10°C   (2) 20°C
(3) 30°C   (4) 60°C

9. Which gas will most closely resemble an ideal gas at STP?
(1) SO_2 (2) NH_3
(3) Cl_2 (4) H_2

10. At standard temperature 1.0 liter of $O_2(g)$ at 760 torr contains the same number of molecules as
   (1) 2.0 L of $O_2(g)$ at 380 torr
   (2) 2.0 L of $O_2(g)$ at 760 torr
   (3) 0.50 L of $O_2(g)$ at 380 torr
   (4) 0.50 L of $O_2(g)$ at 760 torr

11. When the temperature of a sample of water is changed from 17°C to 20°C, the change in its vapor pressure is
   (1) 1.0 mm of Hg   (2) 14.5 mm of Hg
   (3) 3.0 mm of Hg   (4) 17.5 mm of Hg

12. As ice at 0°C changes to water at 0°C, the average kinetic energy of the ice molecules
   (1) decreases (2) increases (3) remains the same

13. A 1 liter flask of $CO_2(g)$ and a 1 liter flask of $H_2(g)$ are both at STP. The ratio of the number of molecules of $CO_2(g)$ to the number of molecules of $H_2(g)$ in the flasks is
   (1) 1:1   (2) 2:3   (3) 3:2   (4) 1:3

14. Which process occurs when dry ice, $CO_2(s)$, is changed into $CO_2(g)$?
   (1) crystallization   (2) condensation
   (3) sublimation   (4) solidification

15. As water in a sealed container is cooled from 20°C to 10°C, its vapor pressure
   (1) decreases   (2) increases   (3) remains the same

16. Water will boil at a temperature of 40°C when the pressure on its surface is
   (1) 14.5 torr   (2) 17.0 torr   (3) 55.3 torr   (4) 760. torr

17. At STP, 44.8 liters of CO_2 contains the same number of molecules as
   (1) 1.00 mole of He   (2) 2.00 moles of Ne
   (3) 0.500 mole of H_2 (4) 4.00 moles of N_2

18. A 1-gram sample of which substance in a sealed 1-liter container will occupy the container completely and uniformly?
   (1) Ag(s)   (2) Hg(l)   (3) $H_2O(l)$ (4) $H_2O(g)$

19. Which Kelvin temperature is equal to −33°C
   (1) −33 K   (2) 33 K   (3) 240 K   (4) 306 K

Base your answers to questions 20 and 21 on the graph below, which represents the uniform heating of a water sample at standard pressure, starting at a temperature below 0°C.

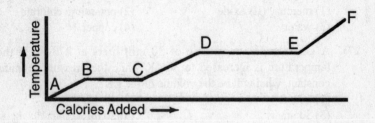

20. The number of calories required to vaporize the entire sample of water at its boiling point is represented by the interval between
(1) *A* and *B*   (2) *E* and *F*   (3) *C* and *D*   (4) *D* and *E*

21. If 5 grams of water undergoes a temperature change from *C* to *D*, the total energy absorbed is
(1) 80 calories   (2) 100 calories   (3) 180 calories   (4) 500 calories

Base your answers to questions 22 and 23 on the graphs shown below.

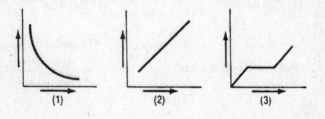

22. Which graph best represents how the volume of a given mass of a gas varies with the Kelvin (absolute) temperature at constant pressure?   (1) 1   (2) 2   (3) 3

23. Which graph best represents how the volume of a given mass of a gas varies with the pressure exerted on it at constant temperature?
(1) 1   (2) 2   (3) 3

24. At STP, which of the following gases has the highest rate of diffusion?   (1) He   (2) Ne   (3) Ar   (4) Kr

25. The vapor pressure of ethanol at its normal boiling point is
(1) 80 mm Hg   (2) 100 mm Hg
(3) 273 mm Hg   (4) 760 mm Hg

26. Which substance *cannot* be decomposed by a chemical change?
(1) mercury (II) oxide   (2) potassium chlorate
(3) water   (4) copper

27. A gas occupies a volume of 30 milliliters at 273 K. If the temperature is increased to 364 K while the pressure remains constant, what will be the volume of the gas?
(1) 60 mL   (2) 40 mL
(3) 30 mL   (4) 20 mL

28. Which quantity represents the total amount of $N_2(g)$ in a 22.4 liter sample at STP?
(1) 1.00 mole   (2) 14.0 grams
(3) 3.01×10^{23} molecules   (4) 6.02×10^{23} atoms

29. Which change of phase represents fusion?
(1) gas to liquid   (2) gas to solid
(3) solid to liquid   (4) liquid to gas

30. Which is the equivalent of 750. calories?
(1) 0.750 kcal   (2) 7.50 kcal
(3) 75.0 kcal   (4) 750. kcal

UNIT 2. ATOMIC STRUCTURE

ATOMS

As early as the 5th century B.C., Greek philosophers believed matter to be composed of separate, indivisible particles that they called atoms. This "particle theory," however, was not generally accepted until the 18th and 19th centuries, when studies of the nature of chemical change helped chemists to develop the atomic theory of the structure of matter.

SUBATOMIC PARTICLES

Most of the volume of an atom consists of empty space. Most of the mass of an atom is concentrated in a dense, centrally located nucleus. The nucleus carries the positive charge of the atom. The region surrounding the nucleus is occupied by tiny, fast-moving, negatively charged particles called electrons.

Nucleons. The particles comprising the nucleus are called nucleons. There are two types of nucleons:

Protons are those particles considered to have a mass of one atomic mass unit and one unit of positive charge.

Neutrons are those particles that have a mass about the same as that of protons, but with a charge of zero.

Electrons. The mass of an electron is about $\frac{1}{1836}$ that of a proton. This is such a small mass that electrons are generally considered to have no mass. An electron has one unit of negative charge that is equal in magnitude (and opposite in electrical nature) to the charge on a proton.

ATOMIC STRUCTURE

The atoms that make up different kinds of matter are different, but their subatomic particles are all alike. Differences in atoms result from differences in the *number* of protons and neutrons in their nuclei and in the number and arrangement of their electrons. **Ernest Rutherford's** experiments in the early 1900's demonstrated that most of the space occupied by an atom is empty most of the time.

Atomic Number. Depending on data collected using X-rays, Henry Moseley showed that each element possessed a characteristic number of positive charge units (protons). This number is now considered to be the **atomic number** of the element, and it is equal to the number of protons in the nucleus.

The charge on the nucleus is often symbolized by the letter "Z." Chemists identify the atomic number as:

- the number of protons in the nucleus;
- the charge on the nucleus;
- or, the Z number.

Atoms of elements are ordinarily electrically neutral, because the number of negative charges (electrons) is equal to the number of positive charges (protons).

Isotopes. Every atom of each element has the same atomic number, because they all have the same number of protons. Some atoms of the same element do, however, have different numbers of neutrons. Since neutrons contribute about the same mass as protons, the masses of atoms of the same element can be slightly different. Atoms of elements having the same atomic number but different masses are called **isotopes**, of that element.

Mass Number. The total number of nucleons (protons + neutrons) in an atom is called the **mass number**, and is characteristic for each isotope. Chemists use the letter "A" as the symbol for mass number, and "N" to designate the number of neutrons. Since the masses of protons and neutrons are each assigned a value of one, the mass number of an atom is a whole number.

The normal designation for isotopes is as follows using sodium -23 as an example:

mass number (protons + neutrons)

$$^{23}_{11}\text{Na}$$

atomic number (protons)

Atomic Mass. The **atomic mass** of an element is the term used to express the weighted average mass of the naturally occurring isotopes of that element. This average is weighted according to the ratios in which the isotopes occur. Most elements occur in nature as mixtures of their isotopes. The mass number of the most abundant isotope can usually be found by rounding off the atomic mass of the element to the nearest whole number.

Electrons. Electrons are located in *energy levels* that exist at various distances outside the nucleus. In neutral atoms, the total number of electrons is equal to the number of protons in the nucleus (the atomic number of the element).

QUESTIONS

1. The number of protons in the nucleus of $^{32}_{15}P$ is   (1)15   (2)17   (3) 32   (4) 47

2. What is the total number of electrons in an atom with an atomic number of 13 and a mass number of 27? (1) 13 (2) 14 (3) 27 (4) 40

3. Atomic mass is measured in atomic mass units (amu) that are based on an atom of
   (1) ^{16}O equal to 16.000 amu   (2) ^{32}S equal to 32.000 amu
   (3) ^{12}C equal to 12.000 amu   (4) ^{14}N equal to 14.000 amu

4. Which symbol represents an isotope of carbon?
   (1) $^{6}_{4}X$ (2) $^{12}_{5}X$ (3) $^{13}_{6}X$ (4) $^{14}_{7}X$

5. The mass number of an atom is equal to the total number of its
   (1) electrons, only (2) protons, only (3) electrons and protons
   (4) protons and neutrons

6. What is the total number of protons in an atom of ^{36}Cl? (1) 17 (2) 18   (3) 35   (4) 36

7. What are the nucleons in an atom? (1) protons and electrons (2) protons and neutrons (3) neutrons and electrons (4) neutrons and positrons

8. Which nuclide contains the greatest number of neutrons?
   (1) ^{37}Cl (2) ^{39}K (3) ^{40}Ar (4) ^{41}Ca

9. In which pair does each atom have the same number of neutrons? (1) ^{24}Mg and ^{24}Na (2) ^{15}N and ^{16}O (3) ^{32}Si and ^{31}P (4) ^{32}S and ^{33}S

10. The neucleus of an atom of $^{127}_{53}I$ contains (1) 53 neutrons and 127 protons (2) 53 protons and 127 neutrons (3) 53 protons and 74 neutrons (4) 53 protons and 74 electrons

11. What is the total number of nucleons (protons and neutrons) in an atom of $^{79}_{34}Se$? (1) 34   (2) 45   (3) 79   (4) 113

12. An atom of $^{226}_{88}$Rn contains (1) 88 protons and 138 neutrons
    (2) 88 protons and 138 electrons (3) 88 electrons and 226 neutrons
    (4) 88 electrons and 226 protons

13. What is the mass number of an atom which contains 21 electrons,
    21 protons, and 24 neutrons? (1) 21   (2) 42   (3) 45   (4) 66

14. A sample of element X contains 90. percent ^{35}X atoms, 8.0 percent
    ^{37}X atoms, and 2.0 percent ^{38}X atoms. The average isotopic mass is
    closest to (1) 32   (2) 35   (3) 37   (4) 38

15. How many electrons are in an atom of beryllium? (1) 5   (2) 2
    (3) 9   (4) 4

16. If X is the symbol of an element, which pair correctly represents
    isotopes of X?

    (1) $^{158}_{64}X$ and $^{158}_{64}X$ (2) $^{64}_{158}X$ and $^{158}_{64}X$ (3) $^{158}_{64}X$ and $^{159}_{64}X$

    (4) $^{158}_{64}X$ and $^{158}_{65}X$

17. Which two particles have approximately the same mass?
    (1) neutron and electron (2) neutron and deuteron (3) proton and
    neutron (4) proton and electron

18. What is the symbol for an atom containing 20 protons and 22
    neutrons? (1) $^{42}_{20}$Ca   (2) $^{40}_{20}$Ca   (3) $^{42}_{22}$Ti   (4) $^{40}_{22}$Ti

MODELS OF ATOMIC STRUCTURE

When Dalton presented his proposal of atomic structure, he pictured
atoms as solid spheres that were indivisible and complete in themselves.
Thomson's discovery of the electron required a new model that included
both positive and negative parts. Thomson proposed an atom that
consisted of a solid, positively charged sphere. The negatively charged
electrons were embedded in the surface of the atom.

Rutherford's experiments showed that the electrons were located
outside the positively charged portion of the atom. His model showed a
dense, positively charged nucleus with electrons somewhere outside the
nucleus. Since Rutherford's time, the model of the atom has undergone
many changes and refinements, and there will probably be many more to
come.

In 1913, the Danish physicist, **Neils Bohr**, proposed a model of the
atom in which electrons revolved around the nucleus in concentric,
circular orbits, like the planets around the sun.

Principal Energy Levels. Each of the orbits in the Bohr model of the atom has a fixed radius. The greater the radius of an orbit (the farther from the nucleus) the greater the energy of the electrons in the orbit. The orbits, or "shells" of the Bohr model are known as **principal energy levels,** and are denoted by the letters K, L, M, N, O, P and Q (or by the numbers 1 through 7).

In this model, when all the electrons are located in the lowest available energy levels, the atom is said to be in the "ground state." However, electrons may absorb a specific amount of energy from outside the atom to move to a higher energy level, leaving a lower level unoccupied. In this condition, the atom is said to be in the "excited state" and is, therefore, unstable. The excited electron(s) will soon "fall" to the lower level. As this occurs, the excess energy is released.

Quanta. One important feature of the Bohr atom is the idea that electrons can only absorb or release energy in discrete, specific, amounts. These amounts, or "bundles" of energy, called **quanta** (singular, quantum), correspond to the differences in energy levels of the shells.

Spectral Lines. If high voltage is applied to hydrogen gas confined in a special tube, called a gas discharge tube, light is emitted. If this light is passed through a prism, a series of bright lines of distinct colors is produced. These lines can be projected onto a screen. Bohr reasoned that these different colored bands of light were actually quanta of correspondingly different energy. These quanta were emitted as electrons of the hydrogen atoms returned from their higher levels in the excited state to their lower levels in the ground state.

The device designed to observe the separation of light into its component colors (wavelengths) is called a **spectroscope**. A spectroscope consists mainly of a prism and a telescope. The series of bright lines produced when excited electrons return to their original energy levels is called a **bright-line spectrum.** Spectral lines can be used to identify elements. The study of these lines has provided much of the evidence regarding energy levels within the atom.

Orbital Model of the Atom

Although Bohr's model of the atom provided an explanation for the bright-line spectrum of hydrogen, it could not account for the spectra of atoms containing many electrons. Scientists, influenced by studies of the wave behavior of electrons, replaced the Bohr model with one that describes the motion of electrons in terms of the *probability* of their positions within the atom. This **orbital model** does not assign specific paths, or orbits, in which the electrons move. Rather, it considers the electrons to move freely around the nucleus. The regions of most probable electron location may differ in size, shape, and orientation in space. These regions are known as **orbitals**.

Energy Levels. In the orbital model, the energy levels of electrons within an atom are represented by quantum numbers. These energy levels can be thought of as having four parts—principal energy levels, sublevels, orbitals, and electron spins.

1. Principal energy levels are similar to the energy levels, or shells, of the Bohr atom. Each principal energy level is represented by a principal quantum number, n. These numbers are 1, 2, 3, 4, 5, 6, and 7. The quantum number is the same as the period number in the periodic table.

2. Sublevels are divisions of principal energy levels. The sublevels help to account for the additional lines present in the spectra of atoms with many electrons. The number of possible sublevels in each principal energy level is equal to the principal quantum number, n. Sublevels are designated by the letters $s, p, d,$ and f.

   a. In describing the energy of an electron in a particular sublevel, both the principal quantum number and one of the letters $(s, p, d,$ or $f)$ are given.

   b. Within each principal energy level, the sublevel of lowest energy is the s sublevel. The other sublevels, in order of increasing energy, are $p, d,$ and f.

   c. The first principal energy level $(n = 1)$ is composed of one sublevel, designated $1s$.

   d. The second principal energy level $(n = 2)$ is composed of 2 sublevels, designated $2s$ and $2p$.

   e. The third principal energy level $(n = 3)$ is composed of 3 sublevels, designated $3s$, $3p$, and $3d$.

   f. The fourth principal energy level $(n = 4)$ is composed of 4 sublevels, designated $4s$, $4p$, $4d$, and $4f$.

3. **Orbitals** are regions within a sublevel in which an electron with a given energy is most likely to be found. Each orbital can hold one or two electrons. The number of orbitals within a given principal energy is determined by squaring the principal quantum number. For example, there are n^2 orbitals in principal energy level n. Orbitals are divided among the sublevels as follows:

   a. In the s sublevel, one orbital is possible.

   b. In the p sublevel, three orbitals are possible.

   c. In the d sublevel, five orbitals are possible.

   d. In the f sublevel, seven orbitals are possible.

4. Electron Spins. Each orbital can contain no more than 2 electrons. (The two electrons in each orbital have opposite spins).

Since each orbital can hold no more than two electrons, it follows that the maximum number of electrons in principal quantum number n is equal to $2n^2$.

This information can be summarized as follows:

Principal Quantum Number (n)	Number of Orbitals (n^2)	Sublevel s p d f	Maximum Number Electrons ($2n^2$)
1	1	1 0 0 0	2
2	4	1 3 0 0	8
3	9	1 3 5 0	18
4	16	1 3 5 7	32

Electron Configurations. There are several different methods for showing, in symbolic form, the distribution of electrons of the atoms of each of the elements in the ground state. These symbolic devices are called **electron configurations**. The following set of rules is the basis for showing these electron distributions:

1. The number of electrons must equal the atomic number.

2. Electrons occupy orbitals in sequence, beginning with those of lowest energy.

3. In a given sublevel, the second electron is not added to an orbital until each orbital in the sublevel contains one electron. (Hund's Rule).

4. No more than four orbitals are occupied in the outermost principal energy level.

In the method known as the "vertical electron configuration," each sublevel is indicated by the principal quantum number and the appropriate letter, s, p, d, or f. The number of electrons present in the orbital or in the sublevel is indicated by a superscript numeral. For example, the notation $3p^3$ indicates that there are three electrons in the $3p$ sublevel.

Electron configurations for the first 18 elements are shown in the following table using two different methods. The Orbital Structure method is more revealing in that it demonstrates **Hund's Rule**. In this method, each box represents an orbital and each ↑↓ a pair of electrons with opposite spins.

TABLE 2-1. ELECTRON CONFIGURATIONS OF THE FIRST 18 ELEMENTS

Element	At.No.	Sublevel Structure	Orbital Structure
			1s · 2s · 2p · 3s · 3p
H	1	$1s^1$	↑
He	2	$1s^2$	↑↓
Li	3	$1s^2 2s^1$	↑↓ · ↑
Be	4	$1s^2 2s^2$	↑↓ · ↑↓
B	5	$1s^2 2s^2 2p^1$	↑↓ · ↑↓ · ↑
C	6	$1s^2 2s^2 2p^2$	↑↓ · ↑↓ · ↑ ↑
N	7	$1s^2 2s^2 2p^3$	↑↓ · ↑↓ · ↑ ↑ ↑
O	8	$1s^2 2s^2 2p^4$	↑↓ · ↑↓ · ↑↓ ↑ ↑
F	9	$1s^2 2s^2 2p^5$	↑↓ · ↑↓ · ↑↓ ↑↓ ↑
Ne	10	$1s^2 2s^2 2p^6$	↑↓ · ↑↓ · ↑↓ ↑↓ ↑↓
Na	11	$1s^2 2s^2 2p^6 3s^1$	↑↓ · ↑↓ · ↑↓ ↑↓ ↑↓ · ↑
Mg	12	$1s^2 2s^2 2p^6 3s^2$	↑↓ · ↑↓ · ↑↓ ↑↓ ↑↓ · ↑↓
Al	13	$1s^2 2s^2 2p^6 3s^2 3p^1$	↑↓ · ↑↓ · ↑↓ ↑↓ ↑↓ · ↑↓ · ↑
Si	14	$1s^2 2s^2 2p^6 3s^2 3p^2$	↑↓ · ↑↓ · ↑↓ ↑↓ ↑↓ · ↑↓ · ↑ ↑
P	15	$1s^2 2s^2 2p^6 3s^2 3p^3$	↑↓ · ↑↓ · ↑↓ ↑↓ ↑↓ · ↑↓ · ↑ ↑ ↑
S	16	$1s^2 2s^2 2p^6 3s^2 3p^4$	↑↓ · ↑↓ · ↑↓ ↑↓ ↑↓ · ↑↓ · ↑↓ ↑ ↑
Cl	17	$1s^2 2s^2 2p^6 3s^2 3p^5$	↑↓ · ↑↓ · ↑↓ ↑↓ ↑↓ · ↑↓ · ↑↓ ↑↓ ↑
Ar	18	$1s^2 2s^2 2p^6 3s^2 3p^6$	↑↓ · ↑↓ · ↑↓ ↑↓ ↑↓ · ↑↓ · ↑↓ ↑↓ ↑↓

Valence Electrons. The electrons that are most responsible for the properties and chemical reactivity of the elements are those in the outermost principal energy level. These electrons are known as **valence electrons**. The rest of the atom—the nucleus and the non-valence electron—is referred to as the "kernel" of the atom.

Sometimes it is useful to show the distribution of valence electrons by means of diagrams called **electron-dot symbols.** In an electron-dot symbol, the kernel of the atom is represented by the chemical symbol of the element. The valence electrons are represented by dots arranged around the kernel. These dots are usually placed in the "clock" positions: 12 o'clock, 3 o'clock, 6 o'clock, and 9 o'clock.

Electron-dot diagrams for the first 18 elements are shown in the following illustration:

$\dot{H}$ $\ddot{He}$

$\dot{Li}$ $\ddot{Be}$ $\dot{B}\cdot$ $\ddot{C}\cdot$ $\cdot\dot{N}\cdot$ $\cdot\ddot{O}:$ $\cdot\ddot{F}:$ $:\ddot{Ne}:$

$\dot{Na}$ $\ddot{Mg}$ $\ddot{Al}\cdot$ $\ddot{Si}\cdot$ $\cdot\dot{P}\cdot$ $\cdot\ddot{S}:$ $\cdot\ddot{Cl}:$ $:\ddot{Ar}:$

Ionization Energy. First Ionization energy is the amount of energy needed to remove the most loosely held electron from an atom of an element in the gas phase. Ionization energies can be found in reference table K and are used to compare the chemical properties and relative activities of the elements. Ionization energies are often expressed in kilocalories per mole of atoms.

More energy is required to remove a second electron from an atom than for the first. This amount of energy is called the *second ionization energy.* Each successive ionization energy is greater than the previous one.

QUESTIONS

1. Which term refers to the region of an atom where an electron is most likely to be found? (1) orbital (2) orbit (3) quantum (4) spectrum

2. An atom has the electron configuration $1s^2 2s^2 2p^6 3s^2 3p^5$. The electron dot symbol for this element is

   (1) $X:$ (2) $\dot{X}:$ (3) $\cdot\dot{X}:$ (4) $:\ddot{X}:$

3. What is the total number of electrons in the second principal energy level of a calcium atom in the ground state? (1) 6 (2) 2 (3) 8 (4) 18

4. Which orbital notation represents the outermost principal energy level of a phosphorus atom in the ground state?

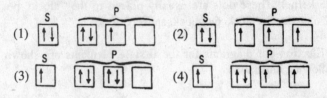

5. Which sublevel has a total of 5 orbitals? (1) s (2) f (3) p (4) d

6. As an electron in an atom falls to a lower energy level, the potential energy of the electron (1) decreases (2) increases (3) remains the same

7. An electron in an atom will emit energy when it moves from sublevel (1) 2s to 3p (2) 2s to 2p (3) 2p to 3s (4) 2p to 1s

8. Which principal energy level change by the electron of a hydrogen atom will cause the greatest amount of energy to be absorbed? (1) $n = 2$ to $n = 4$ (2) $n = 2$ to $n = 5$ (3) $n = 4$ to $n = 2$ (4) $n = 5$ to $n = 2$

9. Spectral lines produced from the radiant energy emitted from excited atoms are thought to be due to the movements of electrons (1) from lower to higher energy levels (2) from higher to lower energy levels (3) in their orbitals (4) out of the nucleus

10. The diagram below represents the orbital notation of an atom's valence shell in the ground state.

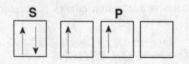

    The diagram could represent the valence shell of (1) Li (2) Si (3) Al (4) Cl

11. The principal quantum number of the outermost electron of an atom in the ground state is $n = 3$. What is the total number of occupied principal energy levels contained in this atom? (1) 1 (2) 2 (3) 3 (4) 4

12. How many occupied sublevels are in an atom of carbon in the ground state? (1) 5 (2) 6 (3) 3 (4) 4

13. What is the total number of valence electrons in an atom with the electron configuration $1s^2 2s^2 2p^6 3s^2 3p^3$? (1) 5 (2) 11 (3) 3 (4) 15

14. The energy level with a principal quantum number *(n)* of 2 contains a total of (1) 8 sublevels (2) 2 sublevels (3) 3 sublevels (4) 4 sublevels

15. In the ground state, how many electrons are in the outermost *s* sublevel of each element in Group 17? (1) 5 (2) 2 (3) 7 (4) 8

16. Usually the term "kernel" includes all parts of the atom *except* the (1) neutrons (2) protons (3) valence electrons (4) orbital electrons

17. Which electron transition is accompanied by the emission of energy? (1) $1s$ to $2s$ (2) $2s$ to $2p$ (3) $3p$ to $3s$ (4) $3p$ to $4p$

18. The maximum number of electrons that a single orbital of the $3d$ sublevel may contain is (1) 5 (2) 2 (3) 3 (4) 4

19. Which is the electron configuration of an atom in the excited state? (1) $1s^2 2s^1$ (2) $1s^2 2s^2 2p^1$ (3) $1s^2 2s^2 2p^5$ (4) $1s^2 2s^2 2p^5 3s^1$

20. The valence electrons represented by the electron dot $\ddot{X}\cdot$ symbol could be those of atoms in Group (1) 13 (IIIA) (2) 15 (VA) (3) 3 (IIIB) (4) 16 (VIA)

21. In an atom, the *s* sublevel has (1) 1 orbital (2) 5 orbitals (3) 3 orbitals (4) 7 orbitals

22. Which electron configuration contains three half-filled orbitals? (1) $1s^2 2s^2 2p^6$ (2) $1s^2 2s^2 2p^5$ (3) $1s^2 2s^2 2p^3$ (4) $1s^2 2s^2 2p^4$

23. What is the first ionization energy of an element that has the electron configuration $1s^2 2s^2 2p^6$? (1) 119 kcal/mol (2) 239 kcal/mol (3) 363 kcal/mol (4) 497 kcal/mol

24. Which is the electron configuration of a neutral atom in the ground state with a total of six valence electrons? (1) $1s^2 2s^2 2p^2$ (2) $1s^2 2s^2 2p^4$ (3) $1s^2 2s^2 2p^6$ (4) $1s^2 2s^2 2p^6 3s^2 3p^6$

25. Which principal energy level has a maximum of three sublevels? (1) 1 (2) 2 (3) 3 (4) 4

26. The total number of orbitals in the $4f$ sublevel is (1) 1 (2) 5 (3) 3 (4) 7

27. Which is the electron dot symbol of an atom of boron in the ground state?

(1) $\cdot\overset{\cdot}{\underset{\cdot}{B}}\colon$ (2) $B\cdot$ (3) $\cdot\overset{\cdot}{\underset{\cdot}{B}}\colon$ (4) $\overset{\cdot}{B}\colon$

28. Which atom has a completely filled 3rd principal energy level? (1) Ar (2) Zn (3) Ca (4) K

29. In which equation does the term "energy" represent the first ionization energy for an atom of potassium?
(1) $K(g) + energy \rightarrow K^+(g) + e^-$ (2) $K(g) + energy \rightarrow K^-(g) + e^-$
(3) $K(g) + e^- \rightarrow K^+ + energy$ (4) $K(g) + e^- \rightarrow K^- + energy$

30. Which is the atomic number of an atom with six valence electrons? (1) 6   (2) 8   (3) 10   (4) 12

31. What is the total number of orbitals in the third principal energy level? (1) 1   (2) 16   (3) 9   (4) 4

NATURAL RADIOACTIVITY

Elements that naturally emit energy without the absorption of energy from an outside source are said to be "naturally radioactive." Elements with atomic number greater than 83 (bismuth) have no known stable isotopes. They are all radioactive. Radioactivity is associated with some form of disintegration, or "decay," of the nucleus. This disintegration results in the emission of particles and/or radiant energy.

When the nucleus of an element disintegrates, it changes to the nucleus of another element. This change from one element to another is called **transmutation.**

Types of Emissions

Soon after the discovery of radioactivity, three different types of emissions were identified. Identification was made on the basis of differences in behavior in an electric field. Those emissions that were attracted to the negative electrode were labeled *alpha* beams. Those attracted to the positive electrode were called *beta* beams. Emissions that were apparently unaffected by charge were called *gamma* beams. It was later found that these emissions also differed from one another in mass, penetrating power, and ionizing power. These findings enabled scientists to identify the three different types of emissions.

Alpha Decay. Alpha particles have been identified as helium nuclei that are ejected at high speeds from certain radioactive isotopes. The reaction involved in nuclear disintegration resulting in the emission of alpha particles is called *alpha decay*. The emitting atoms are called *alpha emitters*. The symbols used to designate isotopes, particles, or radiations in nuclear reactions follow this pattern:

$$_Z^A X$$

in which X is the symbol of the atom or emission, A is the mass number, and Z is the atomic number. (# of protons)

The alpha particle is designated:

$$_2^4 \text{He}$$

An example of an equation for an alpha decay would be:

$$_{88}^{226} \text{Ra} \rightarrow {}_{86}^{222} \text{Rn} + {}_2^4 \text{He}$$

Notice that in nuclear reactions, both mass and charge are balanced. In alpha emission, the isotope's number is reduced by *two,* and its mass number is reduced by *four.*

Beta Decay. Beta particles have the same charge and mass as electrons. Therefore, they are described as electrons ejected at high speeds from certain radioactive isotopes. The reaction is called *beta decay* and the atom is called a *beta emitter.* The equation for the emission of a beta particle from $_{92}^{235}\text{U}$ would be written:

$$_{92}^{235} \text{U} \longrightarrow {}_{-1}^{0} \text{e} + {}_{93}^{235} \text{Np}$$

Since a beta particle has a charge of minus one, balancing the charges brings about an *increase* of one in the atomic number of the product isotope. Beta decay raises at least two questions:

1. How can the product gain a proton without gaining mass?
2. How can the nucleus emit an electron it does not have?

The accepted explanation involves a series of reactions, called *neutron decay,* which results in a neutron in the nucleus being transformed into a proton with the "creation" of an electron.

$$n \rightarrow p + e^-$$

Gamma Radiation. Since gamma rays are not affected by the electric field, they are said to possess zero charge. Since they have no mass, they are not considered particles. Because of their penetrating ability, their properties resemble those of X-rays, but they have much higher energies than most X-rays.

Characteristics of Alpha, Beta and Gamma Emissions

Name	Symbol	Charge	Mass (amu)	Relative Penetrating Power	Relative Ionizing Power
Alpha	$_{2}^{4}He$	+2	4	low	10,000
Beta	$_{-1}^{0}e$	−1	0	moderate	100
Gamma	$_{0}^{0}\gamma$	0	0	high	1

Detection of Radioactivity

The very nature of radioactivity lends itself to various methods of detection and study, both quantitative and qualitative. This "nature" includes its ionizing, fluorescent, and photographic effects.

Half-Life

Radioactive isotopes disintegrate at characteristic rates. The time interval required for half the sample to disintegrate is called the *half- life* of that isotope. The half-life of iodine-131 ($_{53}^{131}I$) is 8 days. If we started with a 100-gram sample of I-131, after 8 days only 50 grams of iodine-131 would remain. After *another* 8 days had passed, only half of the 50 grams, or 25 grams, would remain. After another 8 days, only 12.5 grams would remain, and so on.

Although formulas have been derived for calculating the mass of a radioactive isotope remaining after a given time, it is convenient to set up a simple table of mass and time. For example in the case of iodine-131, how much of a 100-gram sample will remain after 24 days?

Time	Mass
0 days	100g
8 days	50g
16 days	25g
24 days	12.5g (answer)

Remember, it is important to:

1. Find half-life in reference table H if not given.
2. Begin your table with zero time.

QUESTIONS

1. Which element has no known stable isotope?
   (1) carbon   (2) potassium
   (3) polonium   (4) phosphorus

2. Which type of radiation, when passed between two electrically charged plates, would be deflected toward the positive plate?
   (1) an alpha particle   (2) a beta particle
   (3) a neutron   (4) a positron

3. In an electric field, which emanation is deflected toward the negative electrode?
   (1) beta particle   (2) alpha particle
   (3) x rays   (4) gamma rays

4. Which particle is given off when $^{32}_{15}P$ undergoes a transmutation reaction?
   (1) an alpha particle   (2) a beta particle
   (3) a neutron   (4) a positron

5. If 8.0 grams of a sample of ^{60}Co existed in 1990, in which year will the remaining amount of ^{60}Co in the sample be 0.50 gram?
   (1) 1995   (2) 2000   (3) 2006   (4) 2011

6. What is the number of hours required for potassium–42 to undergo 3 half-life periods?
   (1) 6.2 hours   (2) 12.4 hours
   (3) 24.8 hours   (4) 37.2 hours

7. In the equation $^{234}_{91}Pa \rightarrow {}^{234}_{92}U + X$, the X represents an
   (1) helium nucleus   (2) beta particle   (3) proton   (4) neutron

8. In the equation $^{232}_{90}Th \rightarrow {}^{228}_{88}Ra + X$, which particle is represented by the letter X?
   (1) an alpha particle   (2) a beta particle
   (3) a positron   (4) a deuteron

9. Given the reaction:

    $$^{234}_{91}\text{Pa} \rightarrow X + {}^{0}_{-1}\text{e}$$

    When the equation is correctly balanced, the nucleus represented by X is

    (1) $^{234}_{92}\text{U}$ (2) $^{235}_{92}\text{U}$ (3) $^{230}_{90}\text{Th}$ (4) $^{232}_{90}\text{Th}$

10. As an atom of a radioactive isotope emits an alpha particle, the mass number of the atom
    (1) decreases   (2) increases   (3) remains the same

UNIT 3. BONDING

THE NATURE OF CHEMICAL BONDING

The attractive force between atoms or ions is called a **chemical bond.** The nature of this force is often considered basic to the study of chemistry because all chemical changes can be traced to changes in bonds. Normally, within each atom of a given element, the protons of the nucleus and an equal number of electrons outside the nucleus are held together by attractive forces. Under certain conditions, a particular electron (or electron-pair) may become involved with protons of two different nuclei at the same time. Chemical bonds result from this kind of interaction.

Chemical Energy

In Unit I it was stated that energy based on position was called potential energy. The particular form of potential energy involved in the making and breaking of chemical bonds is called **chemical energy.**

Energy Changes in Bonding

Whenever chemical bonds are formed, energy is released (an exothermic process); when bonds are broken, energy is absorbed (an endothermic process). It follows then, that when two atoms are bonded to each other, they are at a lower energy level than when they are alone.

Bonding and Stability

Stability is said to be inversely proportional to potential energy. In other words, systems at lower energy levels are said to be more stable. Whenever large amounts of energy are released in the formation of a bond, the bond is said to be strong and the system very stable. Weak bonds and unstable systems are associated with the release of small amounts of energy. This relationship is very important in the study of kinetics and thermochemistry (Heat of Reaction, Heat of Formation, etc.).

Electronegativity

Electronegativity is the name given to an atom's ability to attract electrons to itself in bond formation. Electronegativity values cannot be measured directly. They are obtained from a variety of data and from a variety of assumptions. The scale of electronegativity values is an arbitrary one. The highest value on the scale is 4.0, assigned to fluorine, the most electronegative of the elements.

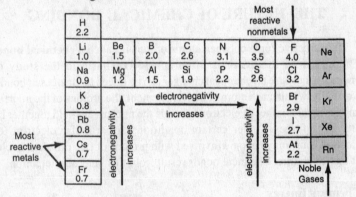

Figure 3-1. Electronegativity Chart

THE BONDS BETWEEN ATOMS

Chemical bonds form when two atoms share electrons or when electrons are transferred from one atom to another. In most cases, the electrons involved in the bonding process are valence electrons. As a result of bond formation, both atoms achieve a valence shell that is identical to one of the noble (inert) gases. This condition—8 valence electrons—is associated with maximum stability and minimum energy content.

Ionic Bonds

Whenever electrons are transferred from one atom to another, the atom that loses electrons becomes a positively charged ion. The atom receiving additional electrons becomes a negatively charged ion. The force of attraction that holds the oppositely charged ions together is called an **ionic bond.**

The ions produced as a result of electron transfer usually have electron configurations identical to those of noble gases. The resulting compounds are said to be ionic. In the solid phase, ionic compounds are held together in very rigid, fixed crystalline structures. Ionic solids are poor conductors of electricity and have high melting points. When ionic solids are melted or dissolved in water, the ions become free to move about and are therefore capable of conducting electricity.

Salts are the most common examples of ionically bonded substances. They can be defined as a metal bonded to a nonmetal (binary salts) or a metal bonded to a polyatomic ion (ternary salts). Examples of binary salts are: NaCl, MgS or $FeBr_2$. Examples of ternary salts are KNO_3 and $Ca(ClO_3)_2$

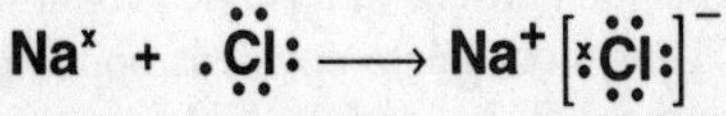

$$Na^x + \overset{\cdot\cdot}{\underset{\cdot\cdot}{Cl}}\overset{\cdot}{:} \longrightarrow Na^+ \left[\overset{\cdot\cdot}{\underset{\cdot\cdot}{:Cl:}}\right]^-$$

Figure 3-2. Sodium chloride is a salt (ionic compound)

Covalent Bonds

When two atoms share electrons, equally or unequally, the resulting bond is called a **covalent bond.** There are several types of covalent bonds.

Nonpolar covalent bonds. Nonpolar covalent bonds form when two atoms with the same electronegativity share electrons. In such cases, electrons are shared equally by the two atoms. Diatomic elements consist of molecules made up of two atoms held together by nonpolar covalent bonds. The chlorine molecule (Cl_2) is one example.

$$\overset{xx}{\underset{xx}{:Cl:}}\overset{\cdot\cdot}{\underset{\cdot\cdot}{Cl:}}$$

Figure 3-3 The Cl_2 molecule has a nonpolar covalent bond.

Polar covalent bonds. Polar covalent bonds form when atoms with different electronegativities share electrons. The polarity is caused by the fact that the electrons are not shared equally. In the hydrogen chloride molecule (HCl), the bond holding the hydrogen and chlorine atoms together is a polar covalent bond.

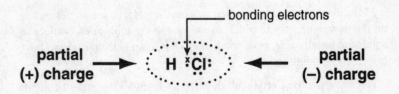

Figure 3-4 The HCl molecule has a polar covalent bond.

Coordinate covalent bonds. Coordinate covalent bonds form when electrons are shared between pairs of atoms in which one of the atoms has no available electron to be shared. In such cases, the other member of the pair contributes both electrons. Once such a bond has formed, it is no different than any other covalent bond.

The coordinate covalent bond is frequently involved in the bonding within polyatomic ions or radicals. The hydrogen ion depends on this type of bonding. This ion consists of a single proton with no electrons to transfer or share. Thus, coordinate covalent bonding is the key to the formation of the ammonium ion and the hydronium ion.

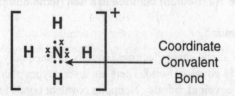

Figure 3-5. The ammonium ion relies on coordinate covalent bonds.

Most bonds have both ionic and covalent characteristics. Comparisons of these characteristics with the difference in the electronegativity values of the reacting species have led to the following conclusions:

1. Whenever the difference in electronegativity values is 1.7 or greater, the bond formed will be predominantly ionic in character.
2. When the electronegativity difference is less than 1.7, the bond formed will be predominantly covalent in character.

Molecular substances. The term molecule is frequently misused. It has been defined as the smallest particle of an element or compound capable of independent motion. While this definition is not incorrect, most chemists reserve the term molecule for particles that consist of covalently bonded atoms, such as H_2, NH_3, H_2O, HCl, CCl_4, and S_8 to name a very few.

Molecular substances may exist as gases, liquids, or solids, depending on the degree of attraction that exists among the molecules. Molecular solids are usually soft, poor conductors of heat and electricity, and have low melting points.

Network solids. Certain solids have an interlocking array of atoms bound together by very strong **covalent bonds**. They are **macromolecules** (giant molecules), and are called **network solids**. Because they are so rigidly held together, network solids are extremely hard. They have very high melting points and are very poor conductors of heat and electricity.

Diamond, the hardest of all naturally occurring substances, is a network solid. In some instances, not all the atoms in a network solid are the same. Silicon carbide (SiC), also known as carborundum, and silicon dioxide (SiO_2) the major component of sand, are both network solids. These substances are widely used for their abrasive, or cutting, qualities. Other examples of network solids are graphite and asbestos.

Metallic bonds. Metals usually have one or two valence electrons. These electrons are readily lost or transferred and, in a metallic solid, they are not always associated with any particular atom. Therefore, the particles in a metal are actually positive ions surrounded by very mobile electrons that can "drift" from one atom to another ("sea" of mobile e^-). The strong bonds that result from the attraction of all the positive ions for the electrons surrounding them are called **metallic bonds**. Their strength accounts for the high melting points of metals. The mobile electrons account for the luster of metals and their ability to conduct heat and electricity.

QUESTIONS

1. Which kind of energy is stored within a chemical substance?
   (1) free energy (2) activation energy (3) kinetic energy
   (4) potential energy
2. Which type of bonding is characteristic of a substance that has a high melting point and electrical conductivity only in the liquid phase? (1) nonpolar covalent (2) coordinate covalent (3) ionic
   (4) metallic
3. Which diagram best represents a polar molecule?

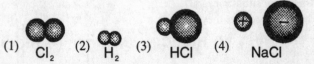

(1) Cl_2 (2) H_2 (3) HCl (4) NaCl

4. Which type of solid does pure water form when it freezes?
   (1) ionic (2) network (3) metallic (4) molecular
5. Atoms of which of the following elements have the strongest attraction for electrons? (1) aluminum (2) chlorine (3) silicon
   (4) sodium
6. Which element is most likely to gain electrons in a chemical reaction? (1) Kr (2) Br (3) Ca (4) Ba
7. Which substance exists as a metallic crystal at STP? (1) Ar (2) Au
   (3) SiO_2 (4) CO_2
8. The P—Cl bond in a molecule of PCl_3 is (1) nonpolar
   (2) polar covalent (3) coordinate covalent (4) electrovalent

9. The electrons in a bond between two iodine atoms (I_2) are shared
    (1) equally, and the resulting bond is polar (2) equally, and the
    resulting bond is nonpolar (3) unequally, and the resulting bond is
    polar (4) unequally, and the resulting bond is nonpolar

10. What type of bond exists in a molecule of hydrogen iodide?
    (1) a polar covalent bond with an electronegativity difference of
    zero (2) a polar covalent bond with an electronegativity
    difference between zero and 1.7 (3) a nonpolar covalent bond with
    an electronegativity difference of zero (4) a nonpolar covalent
    bond with an electronegativity difference between zero and 1.7

11. Given the reaction:

    $$H_2 + Cl_2 \rightarrow 2HCl$$

    Which statement best describes the energy change as bonds are
    formed and broken in this reaction?
    (1) The breaking of the Cl—Cl bond releases energy.
    (2) The breaking of the H—H bond releases energy.
    (3) The forming of the H—Cl bond absorbs energy.
    (4) The forming of the H—Cl bond releases energy.

12. Which atoms are most likely to form covalent bonds (1) metal
    atoms that share electrons (2) metal atoms that share protons
    (3) nonmetal atoms that share electrons (4) nonmetal atoms that
    share protons

13. Which formula represents an ionic compound? (1) KCl (2) HCl
    (3) CO_2 (4) NO_2

14. Which molecule could form a coordinate covalent bond with a
    proton (H^+)?

    (1) H:H   (2) H:C::C:H   (3) H:Ö: (4) H:C:H
                                    H   H
                                                H

15. Which statement correctly describes the bonds in the electron-dot
    formula shown below?

    $$\begin{bmatrix} & H & \\ & \cdot\cdot & \\ H & :N: & H \\ & \times\times & \\ & H & \end{bmatrix}^+$$

    (1) One of the bonds represented is ionic. (2) All of the bonds
    represented are ionic. (3) One of the bonds represented is
    coordinate covalent. (4) All of the bonds represented are
    coordinate covalent.

16. Which element consists of positive ions immersed in a "sea" of mobile electrons? (1) sulfur (2) nitrogen (3) calcium (4) chlorine

17. Which electron-dot diagram best represents a compound that contains both ionic and covalent bonds?

   (1) H:S̈:
        H

   (2) Ca²⁺ [:Ö:S:Ö:]²⁻
              :Ö:

   (3) K⁺[:B̈r̈:]⁻

   (4) :B̈r̈:B̈r̈:

18. Which is the correct electron-dot formula for a molecule of chlorine?

   (1) Cl·Cl   (2) ·C̈l:C̈l·   (3) :C̈l·C̈l: (4) :C̈l:C̈l:

19. The carbon atoms in a diamond are held together by (1) metallic bonds (2) hydrogen bonds (3) ionic bonds (4) covalent bonds

20. The electrical conductivity of KI(aq) is greater than the electrical conductivity of $H_2O(\ell)$ because the KI (aq) contains mobile (1) molecules of H_2O (2) ions from H_2O (3) molecules of KI (4) ions from KI

21. Which compound is an example of a network solid? (1) $H_2O(s)$ (2) $CO_2(s)$ (3) $SiO_2(s)$ (4) $SO_2(s)$

22. Which compound contains both covalent bonds and ionic bonds? (1) NaCl(s) (2) HCl(g) (3) $NaNO_3$(s) (4) N_2O_5(g)

23. Which of the following compounds has the *least* ionic character? (1) KI   (2) NO   (3) HCl   (4) MgS

24. A molecule of ammonia (NH_3) contains (1) ionic bonds, only (2) covalent bonds, only (3) both covalent and ionic bonds (4) neither covalent nor ionic bonds

25. Which bond has the greatest degree of ionic character? (1) Li—Br (2) F—F (3) H—Cl (4) S—O

MOLECULAR ATTRACTION

In addition to the forces responsible for the covalent bonds that hold the atoms together within individual molecules (intramolecular bonds), another set of forces holds the molecules to similar molecules (intermolecular bonds).

Molecules with Dipoles (polar molecules)

The asymmetric, or unequal, distribution of electrical charge in most molecules causes one end of the molecule to be positive and the other end to be negative. This kind of a molecule is a **polar molecule**, which is called a **dipole** (2 poles). The force of attraction between the positive end of one dipole and the negative end of a neighboring dipole helps to hold these molecules together within substances.

Some common examples of asymmetric molecules are H_2S (bent), PH_3 (pyramidal) and HCl (linear).

Molecules with No Dipoles (nonpolar molecules)

Molecules made up of only one kind of atom (H_2, N_2, P_4, S_8) tend to be nonpolar. Electrons are shared equally among all the atoms.

Sometimes the shape of a molecule is such that a uniform distribution of charge results. Such a molecule is nonpolar, even though its individual bonds are polar. For example, CO_2 is a nonpolar molecule held together by polar bonds. The arrangement of the atoms in this molecule can be represented $O = C = O$. The symmetrical shape of the molecule results in an even distribution of charge, making it nonpolar. Another example of a symmetrical molecule is CH_4 (tetrahedral).

Hydrogen Bonding

When hydrogen atoms are covalently bonded to atoms of high electronegativity and small size (oxygen, nitrogen or fluorine), they behave almost as though they were hydrogen ions (bare protons). The reason for this is that they have such a small share of the bonding electron pair. This causes the hydrogen ends of the molecules to be especially attracted to the more electronegative atoms of neighboring molecules. "Super" dipole-dipole attractive forces are known as **hydrogen bonds**. Hydrogen bonds account for some of the special properties of water, such as its unusually high boiling point. Other important molecules with hydrogen bonding are ammonia (NH_3), and hydrogen fluoride (HF).

Van der Waals' Forces

It has been found that even in nonpolar molecules, where there are no dipole attractions or hydrogen bonds, weak attractive forces exist among molecules. These forces have been named van der Waals' forces after the scientist who studied them.

Van der Waals' forces make it possible for small, nonpolar molecules, such as hydrogen, helium, and nitrogen, to exist in the liquid (and even solid) phase under conditions of low temperature and high pressure. This is partially because these forces become more effective as molecules get closer to each other. Van der Waals' forces also increase as the number of electrons increases. Therefore, there is an increase in attraction with increasing molecular size.

Molecule-Ion Attraction

Another form of molecular attraction is the attraction between polar molecules and ions. It is this force that liberates the ions from their crystal lattice in ionic compounds as the compounds are dissolved in polar solvents. For example, when an ionic crystal, such as the salt, sodium chloride (NaCl), is added to water, a polar solvent, the sodium ions (Na^+) are attracted to the negative ends of water molecules. Chloride ions (Cl^-) are attracted to the positive ends of the molecules. The ions are freed from their crystal lattice and are then surrounded by water molecules, forming *hydrated ions*. The orienting of water molecules around ions is called *hydration* of the ions.

Directional Nature of Covalent Bonds

Many properties of compounds can be related to the shapes of their molecules. These shapes, in general, are the result of the directional nature of covalent bonds. A good example here would be the polarity of the water molecule. The molecule is polar because of its bent shape. This asymmetrical shape causes the non-uniform distribution of charge, which results in the polarity.

QUESTIONS

1. Which molecule is a dipole? (1) H_2 (2) N_2 (3) CH_4 (4) HCl
2. Argon has a higher boiling point than neon because argon has (1) fewer electrons in its 2nd principal energy level (2) more electrons in its outermost principal energy level (3) weaker intermolecular forces of attraction (4) stronger intermolecular forces of attraction

3. Which electron dot formula represents a nonpolar molecule?

    (1) H:C:Cl: (2) H:N: (3) H:C:H   (4) H:O:

4. Which of the following liquids has the weakest van der Waal's forces of attraction between its molecules? (1) Xe(ℓ) (2) Kr(ℓ) (3) Ne(ℓ)   (4) He(ℓ)

5. Which molecule is nonpolar and contains a nonpolar covalent bond? (1) CCl_4 (2) F_2 (3) HF   (4) HCl

6. The strongest hydrogen bonds are formed between molecules in which hydrogen is covalently bonded to an element with
    (1) high electronegativity and large atomic radius
    (2) high electronegativity and small atomic radius
    (3) low electronegativity and large atomic radius
    (4) low electronegativity and small atomic radius

7. Van der Waals forces of attraction between molecules always decrease with (1) increasing molecular size and increasing distance between the molecules (2) increasing molecular size and decreasing distance between the molecules (3) decreasing molecular size and increasing distance between the molecules (4) decreasing molecular size and decreasing distance between the molecules

8. In which liquid is hydrogen bonding the most significant force of attraction? (1) HF(ℓ)   (2) HCl(ℓ)   (3) HBr(ℓ)   (4) HI(ℓ)

9. The unusually high boiling point of H_2O is primarily due the presence of (1) hydrogen bonds (2) ionic bonds (3) van der Waals forces (4) molecule-ion attractions

10. At 298 K, the vapor pressure of H_2O is less than the vapor pressure of CS_2. The best explanation for this is that H_2O has (1) larger molecules (2) a larger molecular mass (3) stronger ionic bonds (4) stronger intermolecular forces

11. Which formula represents a polar molecule? (1) CH_4 (2) Cl_2 (3) NH_3 (4) N_2

12. Which formula represents a tetrahedral molecule? (1) CH_4 (2) $CaCl_2$ (3) HBr (4) Br_2

13. Molecule-ion attractions are present in (1) NaCl(aq) (2) HCl(g) (3) CCl_4(ℓ) (4) $KClO_3$(s)

14. As the distance between molecules of a liquid increases, the van der Waals forces of attraction (1) decrease   (2) increase (3) remain the same

CHEMICAL FORMULAS

Chemical formulas use symbols and numerals to provide both qualitative and quantitative information about the composition of substances.

Symbols

Chemical symbols are abbreviations for the elements. Symbols consist of one or two letters. A single-letter symbol is always a capital, or upper-case, letter. In a two-letter symbol, the first letter is always a capital letter. The second letter is always a small, or lower-case, letter.

Formulas

A formula is actually a statement informing us of the identity of the elements present in a substance and the atomic ratios in which they are found in that substance.

Empirical formulas. The formula that depicts the *simplest* atomic ratio in which the elements can combine in a compound is the **empirical formula** of that compound. Formulas for ionic compounds are usually empirical formulas. For example, sodium chloride consists of 1 ion of sodium (Na) for each ion of chlorine (Cl). Its formula is NaCl. Magnesium chloride consists of two ions of chlorine for each ion of magnesium (Mg). Its formula is $MgCl_2$.

Molecular formulas. Covalent structures are comprised of molecules. These molecules sometimes consist of atoms combined chemically in *multiples* of their simplest ratio. In such cases, the empirical formula, while useful, does not tell the whole story. For example, there are many compounds made up of the two elements carbon and hydrogen. The empirical formula for one group of these compounds is CH_2. The ratio of carbon atoms to hydrogen atoms is 1 to 2. Laboratory analysis of a compound from this group may reveal that one molecule of the compound has a mass that is four times the combined masses of one carbon and two hydrogen atoms. Thus, the molecular formula for this particular compound must be 4 times CH_2, or C_4H_8. The molecular formula is sometimes referred to as the "true" formula of the compound.

NAMING AND WRITING FORMULAS
OF CHEMICAL COMPOUNDS

In most cases, chemical formulas are determined by laboratory procedures. Names of compounds are based on the formulas.

Binary compounds are composed of two elements. In binary compounds composed of a metal and a nonmetal, the metallic element is usually named and written first. The name of such compounds ends in -ide. Most compounds of this type are binary *salts*. For example, the compound made up of calcium (metal) and chlorine (nonmetal) has a formula of $CaCl_2$ and is named calcium chloride.

In binary compounds composed of two nonmetals, the less electronegative element is usually named and written first. The name of the compound ends in -ide. Prefixes are used to indicate the number of atoms of each nonmetal. For example, consider two compounds made up of carbon and oxygen. One, having one atom of each element, has the formula CO and is named carbon *mon*oxide. The other, having one carbon atom and two oxygen atoms, has the formula CO_2 and is named carbon dioxide. Some other prefixes used include *tri-* (3), as in sulfur *tri*oxide (SO_3), and *tetra-* (4), as in carbon *tetra*chloride (CCl_4).

Binary acids consist of hydrogen plus some nonmetal. The rule for naming these acids is:

prefix *hydro-* + stem name of nonmetal + suffix -*ic*

Examples:

In HCl, the nonmetal is *chlor*ine, the name of the acid is hydrochloric acid.

In HBr, the nonmetal is *brom*ine, the name of the acid is hydrobromic acid.

Ternary compounds contain three different kinds of atoms. Most compounds that contain three or more different kinds of atoms contain polyatomic ions. Bases consist of metallic ions combined with the hydroxide ion, OH^-. The names of the bases consist of the names of the metallic ion followed by the word *hydroxide*.

Examples:

NaOH is named sodium hydroxide.

$Mg(OH)_2$ is named magnesium hydroxide.

Ternary acids consist of hydrogen ions combined with polyatomic ions. Names of ternary acids are based on the names of the polyatomic ions.

Prefixes and suffixes are applied as needed. If the name of the polyatomic ion ends in *-ite,* the name of the acid ends in *-ous.* If the name of the polyatomic ion ends in *-ate,* the name of the acid ends in *-ic.*

Examples:

Polyatomic ion	Acid	Acid name
$-SO_4$ (sulf*ate*)	H_2SO_4	sulfur*ic* acid
$-SO_3$ (sulf*ite*)	H_2SO_3	sulfur*ous* acid

Names and formulas of polyatomic ions can be found in reference table F.

Stock System. In naming compounds of metals that have more than one oxidation number, the Stock system is used. In this system, the name of the metal is followed by a Roman numeral that represents its oxidation number in that compound. The rest of the name is determined by the appropriate rule.

Examples:

FeO is iron (II) oxide

Fe_2O_3 is iron (III) oxide

$CuSO_4$ is copper (II) sulfate

Cu_2SO_4 is copper (I) sulfate

QUESTIONS

1. Which is a binary compound? (1) potassium hydroxide (2) magnesium sulfate (3) aluminum oxide (4) ammonium chloride

2. The molecular mass of a compound of carbon and hydrogen is 42. Its empirical formula is (1) CH (2) CH_2 (3) CH_3 (4) CH_4

3. Which formula is an empirical formula? (1) H_2CO_3 (2) $H_2C_2O_4$ (3) CH_3COOH (4) CH_2OHCH_2OH

4. Which substance has the same molecular and empirical formulas? (1) C_6H_6 (2) C_2H_4 (3) CH_4 (4) $C_6H_{12}O_6$

5. An example of an empirical formula is (1) C_2H_2 (2) H_2O_2 (3) C_2Cl_2 (4) $CaCl_2$

6. A compound has the empirical formula NO_2. Its molecular formula could be (1) NO_2 (2) N_2O (3) N_4O_2 (4) N_4O_4

7. The formula for calcium cyanide is (1) $CaCN_2$ (2) $CaSCN_2$ (3) $Ca(CN)_2$ (4) $Ca(SCN)_2$

8. Which is the correct formula for titanium (III) oxide? (1) Ti_2O_3
    (2) TiO (3) Ti_3O_2 (4) Ti_2O_4

9. Which is the correct formula for nitrogen (IV) oxide? (1) NO
    (2) NO_2 (3) NO_3 (4) NO_4

10. What is the name for the sodium salt of the acid $HClO_2$?
    (1) sodium chlorite (2) sodium chloride (3) sodium chlorate
    (4) sodium perchlorate

11. What is the name of the calcium salt of sulfuric acid? (1) calcium
    thiosulfate (2) calcium sulfate (3) calcium sulfide (4) calcium
    sulfite

12. What is the correct formula for sodium thiosulfate? (1) $Na_2S_2O_4$
    (2) Na_2SO_2 (3) Na_2SO_4 (4) $Na_2S_2O_3$

13. Which is the formula for magnesium sulfide? (1) MgS (2) $MgSO_3$
    (3) MnS (4) $MnSO_3$

14. A compound has an empirical formula of CH_2 and a molecular
    mass of 56. Its molecular formula is (1) C_2H_4 (2) C_3H_6 (3) C_4H_8
    (4) C_5H_{10}

15. What is the correct formula for chromium (III) oxide? (1) CrO_3
    (2) Cr_3O (3) Cr_2O_3 (4) Cr_3O_2

16. What is the molecular formula of a compound that has a molecular
    mass of 92 and an empirical formula of NO_2? (1) NO_2 (2) N_2O_4
    (3) N_3O_6 (4) N_4O_8

17. What is the correct formula of potassium hydride? (1) KH
    (2) KH_2 (3) KOH (4) $K(OH)_2$

CHEMICAL EQUATIONS

A chemical equation usually represents a chemical reaction. An
equation should identify:

1. the identity of reactants and products
2. the mole ratios of these species
3. some reference to the energy changes involved
4. phases of reactants and products.

When writing chemical equations, the convention is to use an arrow to
indicate the direction of the reaction. (Double arrows are used for

reversible reactions). Reactants are written to the left of the arrow, products to the right of the arrow. Energy is indicated in various ways.

It is important that equations conform to the Law of Conservation of Mass. This means that the total number of atoms of each element present among the products must be the same as that found among the reactants.

In the equation:

$$2H_2\,(g) + O_2(g) \rightarrow 2H_2O(\ell) + heat$$

the following information is given:

- 4 atoms of hydrogen gas react with 2 atoms of oxygen gas.
- 2 molecules of liquid water are produced and heat is released
- 2 moles of hydrogen gas, each consisting of 4 moles of hydrogen atoms, will combine with 1 mole of oxygen gas, consisting of 2 moles of oxygen atoms, to produce 2 moles of liquid water, each consisting of 4 moles of hydrogen atoms and 2 moles of oxygen atoms. Heat energy is also produced by this reaction.

QUESTIONS

1. When the equation $\_\_Al(s) + \_\_O_2(g) \rightarrow \_\_Al_2O_3(s)$ is correctly balanced using the smallest whole numbers, the coefficient of Al(s) is (1) 1   (2) 2   (3) 3   (4) 4

2. Given the unbalanced equation:

    $$\_\_Al_2(SO_4)_3 + \_\_Ca(OH)_2 \rightarrow \_\_Al(OH)_3 + \_\_CaSO_4$$

    When the equation is completely balanced using the smallest whole-number coefficients, the sum of the coefficients is (1) 5   (2) 9   (3) 3   (4) 4

3. Given the equation: $\_\_FeCl_2 + \_\_Na_2CO_3 \rightarrow \_\_FeCO_3 + \_\_NaCl$

    When the equation is correctly balanced using the smallest whole numbers, the coefficient of NaCl is (1) 6   (2) 2   (3) 3   (4) 4

4. Given the balanced equation: $2Na + 2H_2O \rightarrow 2X + H_2$

    What is the correct formula for the product represented by the letter X? (1) NaO (2) Na_2O (3) NaOH   (4) Na_2OH

UNIT 4. THE PERIODIC TABLE

DEVELOPMENT OF THE PERIODIC TABLE

Classification is a very useful device for dealing with information. The periodic table of the elements evolved into its present form as a result of several schemes for classifying the elements. Early in the development of the table, Mendeleev and others considered regularities in properties of the elements to be related to their atomic masses. Moseley has since established that the properties of elements depend on the structure of their atoms and vary with the atomic number in a more systematic way.

Groups. The vertical columns of the periodic table are called **groups** or **families** of elements. At one time they were designated by Roman numerals, but now are more commonly designated by the numerals 1-18. Group 2, for example, consists of Be, Mg, Sr, Ba and Ra. The elements in each group exhibit similar or related properties.

Periods. The horizontal rows of the periodic table are called **periods**, or **rows**. They are designated by the numerals 1-7. Period 3, for example, consists of Na, Mg, Al, Si, P, S, Cl, and Ar. The number of the period is the same as the number of the outermost principal energy level containing electrons. The elements in period 3 have their outermost (valence) electrons in the *third* principal energy level. The properties of the elements in a given period tend to vary in a systematic way.

PROPERTIES OF ELEMENTS IN THE PERIODIC TABLE

Atomic Radius

The radius of an atom is the closest distance to which it will approach another atom of any size under the circumstances specified. The atomic radius is a periodic property of the elements.

Within a single period of the periodic table, the atomic radius generally decreases as the atomic number increases.

The members of any group in the periodic table generally show an increase in atomic radius with an increase in atomic number.

Covalent Radius

The normal covalent radius is the effective distance from the center of the nucleus to the outer valence shell of that atom in a typical covalent or coordinate covalent bond.

Van der Waals Radius

This radius is half the internuclear distance or radius of closest approach of an atom with another atom with which it forms no bond.

Atomic Radius in Metals

This radius is half the internuclear distance or radius in a crystalline metal.

Ionic Radius

Positive ions are produced when metallic atoms lose valence electrons. These ions have smaller radii than do their parent atoms. As you move down a group in the periodic table, ionic radii of positive ions get larger, corresponding to larger numbers of principal energy levels.

Negative ions form when atoms of nonmetals gain electrons in their valence shells. These ions have larger radii than do their parent atoms.

Radii of atoms and of ions are usually expressed in ångstrom units. One ångstrom unit, symbolized 1 Å equals exactly 1×10^{-10} meter.

Metals

Most of the elements (more than two-thirds) are metals. Their atoms lose electrons fairly easily because of their low ionization energies. The resulting ions are positively charged. Metals, therefore, are sometimes referred to as electropositive elements. Their properties include:

1. low ionization energy (energy needed to remove electrons).
2. low electron affinity (attraction for electrons).
3. form positive ions when combining with other atoms.
4. are solids at room temperature (except mercury, Hg, which is a liquid).
5. shine with a metallic luster when polished.
6. are good conductors of heat and electricity.
7. can be rolled or hammered into sheets easily. This property is called *malleability*.
8. can be drawn (pulled) into wire easily. This property is called *ductility*.

Elements with the most pronounced metallic properties appear in the lower left region of the periodic table.

Nonmetals

The atoms of nonmetals gain electrons easily, producing ions with negative charges. They have high ionization energies and high electronegativities. Other properties include:

1. high electron affinities.
2. produce covalent bonds by sharing electrons with other nonmetals.
3. exist as gases, molecular solids, or network solids at room temperature (except bromine, which is a volatile liquid).
4. solids are brittle, possessing little or no malleability or ductility.
5. solids are dull, even when polished.
6. poor conductors of heat and electricity.

The most pronounced nonmetallic elements appear in the upper right region of the periodic table.

Metalloids (semi-metals)

Elements that appear at the border between the metals and the nonmetals have some of the properties of both. They are called *metalloids* or *semi-metals*. Included among the metalloids are the elements boron, silicon, arsenic and tellurium.

QUESTIONS

1. Which element occurs as a solid at STP? (1) bromine (2) carbon (3) mercury (4) nitrogen
2. Which of the following metals has the *lowest* melting point? (1) copper (2) mercury (3) silver (4) iron
3. As the elements in Group 1 of the Periodic Table are considered from top to bottom, each successive element has a (1) smaller first ionization energy (2) larger first ionization energy (3) smaller number of protons (4) larger number of valence electrons
4. Compared to a neon atom, a helium atom has a (1) smaller radius (2) smaller first ionization energy (3) larger atomic number (4) greater number of electrons
5. The radius of a calcium ion is smaller than the radius of a calcium atom because the calcium ion contains the same nuclear charge and (1) fewer protons (2) more protons (3) fewer electrons (4) more electrons
6. Which particle has the largest radius? (1) Cu (2) Cu^{2+} (3) Se (4) Se^{2-}

7. When metal atoms bond with nonmetal atoms, the nonmetal atoms will (1) lose electrons, and the resulting ions are smaller (2) lose electrons, and the resulting ions are larger (3) gain electrons, and the resulting ions are smaller (4) gain electrons, and the resulting ions are larger

8. Compared to atoms of metals, atoms of nonmetals generally have (1) higher electronegativities and lower ionization energies (2) higher electronegativities and higher ionization energies (3) lower electronegativities and lower ionization energies (4) lower electronegativities and higher ionization energies

9. Which element is an active nonmetal? (1) neon (2) oxygen (3) zinc (4) chromium

10. How much energy is required to remove 1 mole of the most loosely bound electrons from 1 mole of $Mg(g)$? (1) 176 kcal (2) 242 kcal (3) 300 kcal (4) 402 kcal

11. Which properties are characteristic of nonmetals? (1) low thermal conductivity and low electrical conductivity (2) low thermal conductivity and high electrical conductivity (3) high thermal conductivity and low electrical conductivity (4) high thermal conductivity and high electrical conductivity

12. The elements in the present Periodic Table are arranged according to their (1) atomic numbers (2) atomic masses (3) mass numbers (4) oxidation states

13. A property of most nonmetals in the solid state is that they are (1) brittle (2) malleable (3) good conductors of electricity (4) good conductors of heat

14. On the Periodic Table, an element classified as a semimetal (metalliod) can be found in (1) Period 6, Group 15, (2) Period 2, Group 14 (3) Period 3, Group 16 (4) Period 4, Group 15

15. Which element listed below has the *least* metallic character? (1) Na (2) Mg (3) Al (4) Si

16. An atom of chlorine and an atom of bromine have the same (1) electronegativity (2) ionization energy (3) van der Waals radius (4) number of valence electrons

THE CHEMISTRY OF A GROUP (FAMILY)

The main reason for the similarity or relatedness of the properties of the elements in a given Group, or Family, with respect to chemical activity is that all members have the same numbers of valence electrons. Typifying this similarity is their behavior as they combine with other elements. The types of compounds produced are very much the same. Group 1 metals form chlorides with the general formula MCl, with M representing any member of the group. The metals in Group 2 form chlorides with the general formula MCl_2. There is a progressive change in the properties of the metals in a group as their atomic number increases. Examples include: the covalent atomic radius, which increases as the atomic number increases; the ionization energy, which generally decreases, because the distance between its nucleus and valence electrons increases. This latter effect is also the result of the larger kernel, containing more filled orbitals. These filled orbitals seem to form a barrier, or "shield," against the attraction between the nucleus and valence electrons. There are some notable exceptions to the rules concerning groups, particularly in Period 2. In period 2, the distance between nucleus and valence electrons is relatively small and there are only two electrons acting as "shields." An example would be boron, the first member of Group 13, which does not produce ions with charge 3+, as do the other members of Group 13.

Groups 1 and 2

Group 1 contains the elements known as the **alkali metals**. Group 2 contains the elements known as the **alkaline earth metals**. These elements have only one or two valence electrons, which are fairly easily lost. They are, therefore, highly reactive. As a result, these elements can be found in nature only in compounds. In most cases, the compounds are processed by electrolytic methods in order to liberate the elemental metals.

Elements in both 1 and 2 exhibit low ionization energies and low electronegativities, and readily form very stable ionic compounds. They do not achieve the electron configuration of the noble gases by covalent bonding. The reactivity of both the alkali metals and the alkaline earth metals increases as we proceed down a group, but this property tends to decrease as the atomic number increases within a given period For example, in period 3, sodium in Group 1 is more reactive than magnesium in Group 2. In this case, the increased reactivity is attributed to the larger atomic sizes in Group 1, when compared to those in Group 2, and also the greater nuclear charge in the elements of Group 2.

Group 15

This group is a rather unique group in that its members range from typical nonmetals (nitrogen and phosphorous) through the metalloids (arsenic and antimony) to the typical metal, bismuth. Reactivity among nonmetals increases as the atomic number *decreases*. Nitrogen and phosphorus are exceptions to this tendency, because they exist as polyatomic molecules.

The relative stability of nitrogen, which is a gas at room temperature, results from the fact that it forms diatomic molecules by means of triple bonds between nitrogen atoms. This means that three pairs of electrons are being shared. It takes a fairly large amount of energy to break such bonds. Nitrogen is a component of proteins, which are absolutely essential for living systems. Unstable compounds of nitrogen are frequently used in producing explosives.

Phosphorus is a component of the nucleic acids, such as DNA and RNA, and is important in the construction of bones and teeth. It is more reactive than nitrogen at room temperature. Phosphorus is normally a solid at room temperature, while nitrogen is a gas. The phosphorus molecule contains four atoms at room temperature. Its greater atomic radius results in bonds that are different than the bonds of nitrogen.

Group 16

Elements in Group 16 show a progression of properties from nonmetallic to metallic as atomic number increases. The first two members of the group, oxygen and sulfur, have typical nonmetallic properties. Selenium and tellurium are metalloids, and the last member, polonium, is a metal. With the exception of oxygen, which is a diatomic gas at room temperature, the elements in Group 16 are solids. As an element, oxygen forms compounds with most other elements fairly easily. In combining with other elements, oxygen exhibits a −2 oxidation state in most cases, and it has a high electronegativity. However, oxygen will combine with fluorine, which is more electronegative. The oxygen behaves like a metal in this case. The abundance of oxygen in the free state in spite of its high reactivity is the result of its being the product of photosynthesis in plants.

The lower reactivity of sulfur accounts for its abundance in its elemental state. This lower reactivity is undoubtedly related to its higher atomic number. Sulfur exhibits both negative and positive oxidation states when it combines with other elements.

Although selenium and tellurium react with hydrogen, with the anticipated negative oxidation state, they usually produce compounds by virtue of positive oxidation states.

The last element in the group, polonium, is highly radioactive. It is produced as a decay product of radioactive uranium. Its radioactivity is associated with the emission of alpha particles, which are helium nuclei. Each alpha emission produces an atom of a stable isotope of lead.

Group 17

The elements of Group 17 are known as the *halogens*. They include fluorine, chlorine, bromine, iodine, and astatine. As with all the nonmetals, the metallic character of the halogens increases with the atomic number. Each of the halogens has a high electronegativity, with fluorine being the most electronegative of *all* the elements. As a result, fluorine exhibits a negative oxidation state in *all* its compounds. The other members of the group demonstrate positive oxidation states when they form compounds with more electronegative elements.

The tendency of halogens to exhibit positive oxidation states when forming compounds increases with increasing atomic number. This is in keeping with the trend in decreased electronegativity as atomic number increases. In the free state, halogens occur as diatomic molecules. At room temperature, fluorine and chlorine are gases, bromine is a liquid, and iodine is a solid. The differences in phase are related to the increasing size of the van der Waal's forces as the size of their molecules increases. Because they are so reactive, it is not surprising to find that halogens occur in nature only as compounds.

The elements themselves are usually extracted from their compounds by removing electrons from their negative ions (halides). Fluorine is the exception because of its extremely high electronegativity. The only way to prepare free fluorine is by melting its compounds, and then submitting the melted compounds to electrolytic processes. Chlorine, bromine, and iodine can be prepared by chemical means.

Group 18

The elements of Group 18 are known as the noble gases. They are also referred to as the inert gases because, it was once thought to be impossible to produce compounds with any of them. However, in recent years, compounds of krypton, xenon, and radon have been produced with the elements fluorine and oxygen. The inability of Group 18 elements to react with other elements is attributed to the stability produced by their electron configurations, in which the outer shells are completely filled.

Transition Elements

The transition elements, which are found in the central region of the periodic table, generally exhibit positive oxidation states. This characteristic is associated with the fact that electrons from an outer s and an inner d sublevel may be involved in chemical reactions. Transition elements produce ions in such a way that their solutions are often variously colored.

QUESTIONS

1. The reactivity of the metals in Groups 1 and 2 generally increases with (1) increased ionization energy (2) increased atomic radius (3) decreased nuclear charge (4) decreased mass

2. The alkaline earth metals are found in Group (1) 1 (IA) (2) 2 (IIA) (3) 11 (IB) (4) 12 (IIB)

3. Which is the most active nonmetallic element in Group 16 (VIA)? (1) oxygen (2) sulfur (3) selenium (4) tellurium

4. Which of the Group 15 (VA) elements can lose an electron most readily? (1) N (2) P (3) Sb (4) Bi

5. When fluorine reacts with a Group 1 metal, it becomes an ion with a charge of (1) 1^- (2) 2^- (3) 1^+ (4) 2^+

6. Which element in Group 15 has the greatest metallic character? (1) N (2) P (3) Sb (4) Bi

7. In which group are all the elements found naturally only in compounds? (1) 18 (2) 2 (3) 11 (4) 14

8. In the ground state, how many electrons are in the outermost s sublevel of each element in Group 17? (1) 5 (2) 2 (3) 7 (4) 8

9. Which three elements have the most similar chemical properties? (1) Ar, Kr, Br (2) K, Rb, Cs (3) B, C, N (4) O, N, Si

10. Which group contains elements in the solid, liquid, and gaseous phase at room temperature? (1) 17 (VIIA) (2) 2 (IIA) (3) 18 (0) (4) 4 (IVB)

11. Which sequence of atomic numbers represents elements which have similar chemical properties? (1) 19, 23, 30, 36 (2) 9, 16, 33, 50 (3) 3, 12, 21, 40 (4) 4, 20, 38, 88

12. In which group are all of the elements solids at STP? (1) 17 (2) 16 (3) 15 (4) 14

13. Which groups contain metals that are so active chemically that they occur naturally only in compounds?
(1) 1 (IA) and 2 (IIA) (2) 2 (IIA) and 12 (IIB)
(3) 1 (IA) and 11 (IB) (4) 11 (IB) and 12 (IIB)

14. In which Group do the elements usually form oxides which have the general formula M_2O_3? (1) 1 (2) 2 (3) 13 (4) 14

15. Element X forms the compounds XCl_3 and X_2O_3. In the Periodic Table, element X would most likely be found in Group (1) 1 (IA) (2) 2 (IIA) (3) 13 (IIIA) (4) 14 (IVA)

16. Which halogen has the least attraction for electrons? (1) F (2) Cl (3) Br (4) I

17. As the elements in Group 2 are considered from beryllium to radium, the degree of metallic activity
    (1) increases and atomic radius increases
    (2) increases and atomic radius decreases
    (3) decreases and atomic radius increases
    (4) decreases and atomic radius decreases

18. Which of the following Group 18 elements would be most likely to form a compound with fluorine? (1) He (2) Ne (3) Ar (4) Kr

19. Which group of elements in the Periodic Table contains a semimetal (metalloid)? (1) 1 (IA) (2) 7 (VIIB) (3) 13 (IIIA) (4) 18 (0)

20. Which element at STP exists as monatomic molecules? (1) N (2) O (3) Cl (4) Ne

21. Oxygen has a positive oxidation number when it is in a compound with (1) I (2) Br (3) Cl (4) F

22. Which element exhibits a crystalline structure at STP? (1) fluorine (2) chlorine (3) bromine (4) iodine

23. Which group contains elements with a total of four electrons in the outermost principal energy level? (1) 1 (2) 18 (3) 16 (4) 14

24. Which of the following Group 17 (VIIA) elements has the highest melting point? (1) fluorine (2) chlorine (3) bromine (4) iodine

25. Which element is a liquid at room temperature (1) K (2) I_2 (3) Br_2 (4) Mg

26. As the elements in Group 18 are considered in order of increasing atomic number, the ionization energy of each successive element (1) decreases (2) increases (3) remains the same

27. Elements whose two outermost sublevels may be involved in a chemical reaction are called (1) noble gases (2) halogens (3) alkali metals (4) transition metals

28. Which atom has multiple oxidation states and forms an ion that is colored when in solution? (1) Cl (2) F (3) Cu (4) Zn

29. Which element may react chemically by losing electrons from both s and d sublevels? (1) Al (2) Cl (3) Fe (4) Be

30. The water solution of a compound is bright yellow. The compound could be (1) KNO_2 (2) K_2CrO_4 (3) KOH (4) K_3PO_4

31. Which compound is colorless in a water solution? (1) $Al_2(SO_4)_3$ (2) $Cr_2(SO_4)_3$ (3) $Fe_2(SO_4)_3$ (4) $Co_2(SO_4)_3$

32. A white anhydrous powder that dissolves in water to form a blue aqueous solution could be (1) $MgSO_4$ (2) $BaSO_4$ (3) $CuSO_4$ (4) $CaSO_4$

33. Which of the following aqueous solutions is blue? (1) $Na_2SO_4(aq)$ (2) $K_2SO_4(aq)$ (3) $MgSO_4(aq)$ (4) $CuSO_4(aq)$

34. As the atoms of the elements in Group I (IA) are considered in order from top to bottom, compared to the ionization energy of the atom above it, the ionization energy of each successive atom (1) decreases (2) increases (3) remains the same

THE CHEMISTRY OF A PERIOD

In each period of the periodic table, the elements have the same number of principal energy levels. Their valence electrons are in the same principal energy level. In a given period, the properties of the elements change systematically as their atomic numbers increase.

As the atomic number increases:

1. the covalent atomic radius decreases.

2. there is an increase in the ionization energy.

3. electronegativity increases.

4. there is a general change in character from that of very active metallic elements to that of the less active nonmetallic elements, to very active nonmetals, and, finally, to that of noble gases.

5. there is a gradual change from positive to negative oxidation states.

6. the metallic characteristics of the elements decrease. It should be pointed out that this trend does not apply to the transition elements.

QUESTIONS

1. Of all the elements, the one with the highest electronegativity is found in Period (1) 1 (2) 2 (3) 3 (4) 4

2. Which element exists as a diatomic molecule at STP? (1) bromine (2) argon (3) sulfur (4) rubidium

3. As the elements of Period 2 are considered in succession from left to right, there is a general decrease in (1) ionization energy (2) electronegativity (3) metallic character (4) nonmetallic character

4. Which element in Period 3 has the most metallic character? (1) Al (2) Si (3) Na (4) Mg

5. Which element in Period 4 of the Periodic Table exhibits the most nonmetallic properties? (1) Ca (2) Cr (3) Ga (4) Br

6. In which of the following periods of the periodic Table are transition elements found? (1) 1 (2) 2 (3) 3 (4) 4

7. The highest ionization energies in any period are found in Group
    (1) 1 (IA)   (2) 2 (IIA)   (3) 17 (VIIA)   (4) 18 (O)

8. Which element in Period 2 has the greatest tendency to gain
    electrons? (1) Li   (2) C   (3) F   (4) Ne

9. Elements in Period 3 are alike in that they all have the same
    number of (1) protons (2) neutrons (3) electrons in the valence
    shell (4) occupied principal energy levels

10. Which element in Period 3 is the most active nonmetal?
    (1) sodium   (2) magnesium   (3) chlorine   (4) argon

11. Which electron configurations represent the first two elements in
    Group 17 (VIIA) of the Periodic Table?
    (1) $1s^22s^1$ and $1s^22s^2$ (2) $1s^22s^2$ and $1s^22s^22p^1$
    (3) $1s^22s^22p^5$ and $[Ne]3s^23p^5$ (4) $1s^22s^22p^6$ and $[Ne]3s^23p^5$

12. - 15. Answer on the basis of the following hypothetical periodic table:

Group↔ / Period	1	2	13	14	15	16	17	18
2	I			H		G		A
3	B	C	D	E	F		M	
4	J		K		L			

12. An element with a partially full **p** orbital is (1) A (2) B
    (3) G   (4) I

13. Which two elements combine to form the compound with the
    highest percent ionic character? (1) A and G (2) B and F
    (3) B and G   (4) H and E

14. The ionization potential for the first electron will be greatest for
    element (1) A   (2) B   (3) E   (4) G

15. Two elements with similar orbital structures are (1) A and I
    (2) B and C   (3) B and J   (4) E and F

16. Which element within any given period of the Periodic Table would
    always have the *lowest* first ionization energy? (1) an alkali metal
    (2) an alkaline earth metal (3) a halogen (4) a noble gas

17. When an atom of chlorine forms an ionic bond with an atom of sodium, the atom of chlorine (1) loses an electron (2) loses a proton (3) becomes an ion with a smaller radius than the atom of chlorine (4) becomes an ion with a larger radius than the atom of chlorine

18. Which of the following electron configurations represents the element with the *smallest* covalent radius? (1) $1s^2 2s^2 2p^2$ (2) $1s^2 2s^2 2p^3$ (3) $1s^2 2s^2 2p^4$ (4) $1s^2 2s^2 2p^5$

UNIT 5. MATHEMATICS OF CHEMISTRY

MOLE INTERPRETATION

Since finding the mass of a single atom would be virtually impossible, some standard unit is needed to compare conveniently the masses in grams of the atoms of elements or compounds. The term mole is now used to name this number in the same way a dozen is used to name the number 12 or a gross is used to name the number 144.

The term *mole* was selected to indicate the number of molecules in a molecular mass of a compound, expressed in grams, or the number of atoms in the atomic mass of an element, expressed in grams. The number has since been found to be 6.02×10^{23}. It has been named Avogadro's number in honor of the Italian professor of physics.

USE OF THE MOLE CONCEPT

Gram atomic mass (gram-atom)

The mass of 6.02×10^{23} atoms, or 1 mole of atoms, of an element represents the **gram atomic mass** of that element. It is also called one **gram-atom** of the element. This quantity is numerically equal to the atomic mass of the element, which can be found in the periodic table.

Gram Molecular Mass

The mass of 6.02×10^{23} molecules, or 1 mole of molecules, of a substance represents the **gram molecular mass,** or **mole mass** of that substance. It is found by adding the atomic masses of all the atoms in a given molecule of the substance.

The sum of the masses of ions, or one mole of empirical units, of an ionic substance is called the **formula mass** of that substance. This mass is used in calculations dealing with ionic substances and also network solids, since they are not composed of molecules.

Molar Volume of a Gas.

The volume occupied by one mole of a substance is known as the **molar volume** of that substance. The molar volume is the same for all gases at the same temperature and pressure. At STP, the molar volume of any gas is 22.4 liters.

SAMPLE PROBLEMS

Find the gram formula mass for sodium sulfate.

Solution:

1. Write the correct formula Na_2SO_4
2. Find the gram atomic mass of each element in the periodic table.
3. Multiply each atomic mass by the subscript in the formula.
4. Find the sum of all the masses.

Element	g-atomic Mass		Subscript	Totals
Na	23g	×	2	=46g of Na
S	32g	×	1	=32g of S
O	16g	×	4	=64g of O
One mole of Na_2SO_4				**=142g of Na_2SO_4**

Find the mass of one mole of glucose, $C_6H_{12}O_6$.

C = 12g × 6 = 72g C
H = 1g × 12 = 12g H
O = 16g × 6 = 96g O
Answer = 180g $C_6H_{12}O_6$

QUESTIONS

1. What is the formula mass of $Al_2(SO_4)_3$? (1) 123 (2) 150. (3) 214 (4) 342

2. Which represents the greatest mass of chlorine? (1) 1 mole of chlorine (2) 1 atom of chlorine (3) 1 gram of chlorine (4) 1 molecule of chlorine

3. What is the total number of hydrogen atoms in one mole of CH_3OH? (1) 6.0×10^{23} (2) 18×10^{23} (3) 24×10^{23} (4) 36×10^{23}

4. If a gas has a molecular mass of 44.0, the volume of 88.0 grams of this gas at STP would be (1) 11.2 L (2) 22.0 L (3) 44.8L (4) 88.0 L

5. What is the total number of atoms of oxygen in the formula $Al(ClO_3)_3 \cdot 6H_2O$? (1) 6 (2) 9 (3) 10 (4) 15

6. What is the total number of moles of oxygen atoms in 1 mole of ozone? (1) 1 (2) 2 (3) 3 (4) 4

7. What is the total number of moles of atoms present in 1 mole of
    $Ca_3(PO_4)_2$?   (1) 13   (2) 10   (3) 8   (4) 5
8. In the compound Al_2O_3, the ratio of aluminum to oxygen is (1) 2
    grams of aluminum to 3 grams of oxygen   (2) 3 grams of
    aluminum to 2 grams of oxygen   (3) 2 moles of aluminum to 3
    moles of oxygen   (4) 3 moles of aluminum to 2 moles of oxygen
9. How many molecules are in 0.25 mole of O_2?   (1) 12×10^{23}
    (2) 6.0×10^{23} (3) 3.0×10^{23} (4) 1.5×10^{23}
10. Which gas sample contains a total of 3.0×10^{23} molecules?
    (1) 71 g of Cl_2 (2) 2.0 g of H_2 (3) 14 g of N_2 (4) 38 g of F_2
11. At STP, 5.6 liters of CH_4 contains the same number of molecules
    as (1) 1.4 L of oxygen   (2) 2.8 L of ammonia   (3) 5.6 L of
    hydrogen (4) 11.2 L of neon
12. What mass of carbon dioxide occupies a volume of 22.4 liters at
    STP? (1) 22.0 g (2) 44.0 g (3) 66.0 g (4) 88.0 g

STOICHIOMETRY

Calculations based on quantitative relationships in a balanced
chemical equation are referred to as stoichiometry. This term comes from
the Greek words *stoicheon* (element) and *metrein* (to measure).

In order to use stoichiometric methods, we must assume that:
1. the reaction has no side reactions.
2. reaction goes to completion.
3. the reactants are completely consumed.

Problems Involving Formulas

Percent Composition. Since the formula of a compound provides the
mole ratios of its component atoms, it is possible to determine the
number of moles of each element present in a given mass of the
compound. With this information, the percentage, by mass, of each
element can then be calculated.

SAMPLE PROBLEM

Determine the percent composition of mercuric (II) chloride, $HgCl_2$.

Solution:

Step 1. Calculate the formula mass of $HgCl_2$:

$$201 + (2 \times 35.5) = 272$$

Step 2. Calculate the percentage of the formula mass due to each component:

$$\frac{\text{mass due to Hg}}{\text{formula mass}} = \frac{201}{272} = 0.739 \text{ or } \textbf{73.9\% Hg}$$

$$\frac{\text{mass due to Cl}_2}{\text{formula mass}} = \frac{71.0}{272} = 0.261 \text{ or } \textbf{26.1\% Cl}$$

Check: $73.9\% + 26.1\% = 100.0\%$

Empirical Formula. The simplest mole ratio among the component elements in a compound is represented by whole-number subscripts in the empirical formula for that compound. The molecular, or "true," formula for the compound can be any simple multiple of the empirical formula, depending on the molar mass of the compound.

Consider the group of compounds made up of atoms of carbon and hydrogen. For the compound that contains 2 moles of hydrogen for each mole of carbon, the empirical formula is CH_2. The empirical formula mass of this compound is:

$$
\begin{aligned}
C &= 12g \times 1 = 12g \quad \text{carbon} \\
H &= 1g \times 2 = \underline{2g} \quad \text{hydrogen} \\
&\qquad\qquad\quad 14g \ CH_2
\end{aligned}
$$

If laboratory analysis reveals that the gram molecular mass of a compound is actually 14g, then CH_2 is the molecular, as well as the empirical, formula of the compound. However, suppose the gram molecular mass is found to be 42g/mole. Since $42 \div 14 = 3$, the molecular formula for the compound must be $3(CH_2)$, or C_3H_6.

Empirical Formula from Percent Composition

Find the empirical formula of a compound composed of 75% carbon and 25% hydrogen by mass.

Solution:

1. Assume a 100-gram sample.

2. Find the mass of each element in the sample:
   mass of C = 75% of 100 grams = 75g C
   mass of H = 25% of 100 grams = 25g H

3. Convert grams to moles:

$$75g\ C\ \times \frac{1\ mole\ C}{12g\ C} = 6.25\ moles\ C$$

$$25g\ H\ \times \frac{1\ mole\ H}{1.0\ g\ H} = 25\ moles\ H$$

4. Find the mole ratio (divide both numbers by the smaller number):

   $6.25 \div 6.25 = 1$ C

   $25 \div 6.25 = 4$ H

   empirical formula = CH_4

Molecular Formula from Empirical Formula and Gram Molecular Mass

A compound with empirical formula CH_3 is found to have a gram molecular mass of 30 grams. Find its molecular formula.

Solution:

1. Find the empirical formula mass of CH_3:

   C = 12g × 1 = 12g C
   H = 1g × 3 = 3g H
   15g = empirical formula mass of CH_3

2. Find the molecular formula:

$$\text{molecular formula} = \frac{\text{empirical}}{\text{formula}} \times \frac{\text{molecular mass}}{\text{empirical formula mass}}$$

$$= CH_3 \times \frac{30g}{15g}$$

$$= CH_3 \times 2 = C_2H_6$$

molecular formula $= C_2H_6$

Molecular Mass of Gas from Mass Density

Find the molecular mass of a gas having a density of 1.96 grams per liter at STP.

Solution:

The volume of one mole of any gas at STP is 22.4 liters.

$$\text{Density} = \frac{\text{mass}}{\text{volume}} \quad \text{or} \quad \text{mass} = \text{Density} \times \text{volume}$$

Thus: mass of 1 mole $= \dfrac{1.96 \text{ g}}{1 \text{ liter}} \times 22.4$ liters $= 43.9$ grams

molecular mass $= \textbf{43.9 grams}$

Problems Involving Equations

   Mass problems. A balanced equation provides a considerable amount of information about the reaction it represents. It not only identifies the reactants and products, it also gives the mole proportions of all the substances involved in the reaction. These proportions are shown by the coefficients used to balance the equation. Consider the following:

$$2Mg(s) + O_2(g) \rightarrow 2MgO(s)$$

This equation tells us that 2 moles of Mg metal react with 1 mole of oxygen gas to produce 2 moles of magnesium oxide (a white powder). Using this information, it is possible to calculate the mass of any substance involved in this reaction given the mass of any one of the other substances involved.

SAMPLE PROBLEM

How many grams of magnesium are needed to produce 120 grams of magnesium oxide in the reaction:

$$2Mg(s) + O_2(g) \rightarrow 2MgO(s)$$

Solution:

1. Identify the known quantity and the unknown quantity to be calculated:

   Known = 120g MgO
   Unknown = xg Mg

2. Plan the sequence of conversions needed to proceed from the known quantity to the unknown quantity:

$$2\,Mg + O_2 \longrightarrow 2\,MgO$$
   grams Mg   120 g MgO

        ↑   ↓

   moles Mg ⟵————— moles MgO

3. Using conversion factors (moles $\rightleftarrows$ g) and mole ratio (from the equation), set up and perform the operations:

$$120 \text{ g MgO} \times \frac{1 \text{ mole MgO}}{40. \text{ g MgO}} \times \frac{2 \text{ moles Mg}}{2 \text{ moles MgO}} \times \frac{24 \text{ g Mg}}{1 \text{ mole Mg}}$$

= 72 grams Mg

Mass-Volume Problems

This method of solving quantitative problems based on the mole concept can also be used when the problem involves the mass of one substance (usually in grams) and the volume of another substance (usually in liters). The mole unit can be used to relate quantities of reactants and products. At STP, the molar volume for all gases is 22.4 liters.

SAMPLE PROBLEM

For the following reaction: at STP:

$$Zn(s) + 2HCl(aq) \rightarrow ZnCl_2(aq) + H_2(g)$$

how many grams of zinc are needed to produce 11.2 liters of hydrogen gas?

Solution:

1. Identify the known quantity and unknown quantity to be calculated:

   Known   11.2L H_2 gas
   Unknown   Xg Zn

2. $Zn + 2HCl \longrightarrow ZnCl_2 + H_2$
   x g Zn   11.2 L H_2
     ↑   ↓
   moles Zn ←——————— moles H_2

3. $11.2L\ H_2 \times \dfrac{1\ mole\ H_2}{22.4L\ H_2} \times \dfrac{65.4g\ Zn}{1\ mole\ Zn} = \textbf{32.5 grams Zn}$

Volume-Volume Problems

In the case of reactions involving only gases, calculations are relatively simple. This is because, at the same temperature and pressure, mole ratios and volume ratios are the same for all gases. The coefficients in the balanced equation represent not only the mole ratios, but the volume ratios of reactants and products involved in the reaction.

SAMPLE PROBLEM

Given the reaction

$$3H_2(g) + N_2(g) \rightarrow 2NH_3(g)$$

find the volume of hydrogen gas needed to produce 100. liters of ammonia gas at STP.

Solution:

$$100.\text{L NH}_3 \times \frac{3\text{L H}_2}{2\text{L NH}_3} = 150.\text{ liters H}_2$$

QUESTIONS

1. The approximate percent by mass of potassium in $KHCO_3$ is
   (1) 19% (2) 24% (3) 39% (4) 61%

2. The percent by mass of nitrogen in NH_4NO_3 (formula mass=80) is approximately (1) 18% (2) 23% (3) 32% (4) 35%

3. What is the empirical formula of a compound consisting of 29.6% oxygen and 70.4% fluorine by mass? (1) OF (2) OF_2 (3) O_2F (4) O_2F_4

4. Vitamin C has an empirical formula of $C_3H_4O_3$ and a molecular mass of 176. What is the molecular formula of vitamin C?
   (1) $C_3H_4O_3$ (2) $C_6H_8O_6$ (3) $C_9H_{12}O_9$ (4) $C_{10}H_8O_3$

5. A compound has an empirical formula of CH_2 and a molecular mass of 56. What is its molecular formula? (1) CH_2 (2) C_2H_4 (3) C_3H_6 (4) C_4H_8

6. The empirical formula of a compound is CH_2O and the molecular mass is 180. What is the molecular formula of this compound?
   (1) $C_6H_{12}O_6$ (2) $C_4H_8O_4$ (3) $C_2H_4O_2$ (4) CH_2O

7. Given the reaction:

   $$2C_2H_6 + 7O_2 \rightarrow 4CO_2 + 6H_2O$$

   What is the total number of CO_2 molecules produced when one mole of C_2H_6 is consumed? (1) 6.02×10^{23} (2) $2(6.02 \times 10^{23})$ (3) $3(6.02 \times 10^{23})$ (4) $4(6.02 \times 10^{23})$

8. The density of a gas is 0.77 gram per liter at STP. What is the formula mass of the gas? (1) 8.5 g (2) 17 g (3) 29 g (4) 34 g

9. Given the reaction:

$$(NH_4)_2CO_3 \rightarrow 2NH_3 + CO_2 + H_2O$$

What is the minimum amount of ammonium carbonate that reacts to produce 1.0 mole of ammonia? (1) 0.25 mole (2) 0.50 mole (3) 17 moles (4) 34 moles

10. Given the reaction:

$$4Na + O_2 \rightarrow 2Na_2O$$

How many grams of oxygen are completely consumed in production of 1.00 mole of Na_2O? (1) 16.0 (2) 32.0 (3) 62.0 (4) 124

11. Given the balanced equation:

$$NaOH + HCl \rightarrow NaCl + H_2O$$

What is the total number of grams of H_2O produced when 116 grams of the product, NaCl, is formed? (1) 9.0 g (2) 18 g (3) 36 g (4) 54 g

12. Given the reaction:

$$4Al + 3O_2 \rightarrow 2Al_2O_3$$

How many moles of Al_2O_3 will be formed when 27 grams of Al reacts completely with O_2? (1) 1.0 (2) 2.0 (3) 0.50 (4) 4.0

13. The gram molecular mass of a gas is 44.0 grams. What is its density at STP? (1) 0.509 g/L (2) 1.43 g/L (3) 1.96 g/L (4) 2.84g/L

14. Given the reaction:

$$C_3H_8(g) + 5O_2(g) \rightarrow 3CO_2(g) + 4H_2O(g)$$

What is the total volume of $H_2O(g)$ formed when 8.00 liters of $C_3H_8(g)$ is completely oxidized? (1) 32.0 L (2) 22.4 L (3) 8.00 L (4) 4.00 L

15. Given the reaction:

$$H_2(g) + Cl_2(g) \rightarrow 2HCl(g)$$

What is the total volume of H_2 gas consumed when 22.4 liters of Cl_2 gas completely reacts? (1) 11.2 L (2) 22.4 L (3) 44.8 L (4) 89.6 L

16. According to the reaction $H_2 + Cl_2 \rightarrow 2HCl$, the production of 2.0 moles of HCl would require 70. grams of Cl_2 and (1) 1.0 g of H_2 (2) 2.0 g of H_2 (3) 3.0 g of H_2 (4) 4.0 g of H_2

17. Given the reaction:

$$4NH_3(g) + 5O_2(g) \rightarrow 4NO(g) + 6H_2O(g)$$

What is the total number of liters of $O_2(g)$ required to produce 40 liters of NO? (1) 5 L (2) 9 L (3) 32 L (4) 50 L

18. What is the empirical formula of a compound whose composition by mass is 40.% sulfur and 60.% oxygen? (1) SO_2 (2) SO_3 (3) S_2O_3 (4) S_2O_7

19. Given the reaction:

$$2Na + 2H_2O \rightarrow 2NaOH + H_2$$

How many moles of Na are needed to produce exactly 5.6 liters of H_2, measured at STP? (1) 1.0 (2) 2.0 (3) 0.50 (4) 0.25

20. How many moles of water are contained in 0.250 mole of $CuSO_4 \bullet 5H_2O$? (1) 1.25 (2) 4.50 (3) 40.0 (4) 62.5

21. A 16.0-gram sample of a compound containing only copper and sulfur is decomposed to produce 12.8 grams of copper and 3.2 grams of sulfur. The empirical formula of the compound is (1) CuS (2) Cu_2S (3) CuS_2 (4) Cu_2S_3

22. Which is an empirical formula? (1) C_4H_8 (2) C_2H_2 (3) H_2O (4) H_2O_2

23. What is the percent by mass of hydrogen in CH_3COOH (formula mass = 60.)? (1) 1.7% (2) 5.0% (3) 6.7% (4) 7.1%

24. Given the reaction:

$$2CH_3OH(\ell) + 3O_2(g) \rightarrow 2CO_2(g) + 4H_2O(g)$$

How many liters of $O_2(g)$ are needed to produce exactly 200 liters of $CO_2(g)$? (1) 100 L (2) 200 L (3) 300 L (4) 400 L

25. A compound consists of 46.7% nitrogen and 53.3% oxygen by mass. What is its empirical formula? (1) NO (2) NO_2 (3) N_2O (4) N_2O_3

26. The percent by mass of nitrogen in $Mg(CN)_2$ is equal to

(1) $\dfrac{14}{76} \times 100$ (2) $\dfrac{14}{50} \times 100$ (3) $\dfrac{28}{76} \times 100$ (4) $\dfrac{28}{50} \times 100$

27. Given the balanced equation:

$$3Fe + 4H_2O \rightarrow Fe_3O_4 + 4H_2$$

What is the total number of liters of H_2 produced at STP when 36.0 grams of H_2O is consumed? (1) 22.4 (2) 33.6 (3) 44.8 (4) 89.6

SOLUTIONS

A solution is defined as a homogeneous mixture of two or more substances. The substance that is dissolved is called the **solute.** The substance in which the solute is dissolved is called the **solvent.**

In beginning chemistry courses, most solutions have water as the solvent. All such solutions are called **aqueous solutions**.

The most common solutions consist of solid solutes dissolved in liquid solvents. Solutions consisting of two liquids are also fairly common. In such solutions, the liquid present in excess is considered to be the solvent.

Some liquids cannot be mixed together to form solutions. Such liquids are said to be *immiscible.* A mixture of oil and water is an example of immiscible liquids. Liquids that will form homogeneous mixtures in almost any proportions, such as methyl alcohol and ethyl alcohol, are said to be *miscible.* Mixtures of miscible liquids are true solutions.

Methods of Indicating Solubility of Solutes

The **solubility** of a substance in a given solvent refers to how much of the substance can be dissolved in a given quantity of the solvent at a specified temperature and pressure. Words like *soluble* and *insoluble* are used to indicate that relatively very large or very small amounts of the solute will dissolve. Certain tables use specified terms to describe *ranges* of solubility. Terms like nearly insoluble, slightly soluble, soluble, and very soluble are often used.

Sometimes solubility curves are provided in which the maximum quantity of solute which will dissolve in 100g of solvent is plotted against temperature.

Saturated solution is the expression used when these conditions prevail. That is, a saturated solution contains the maximum solute that will dissolve in a given quantity of solvent at a specified temperature.

Unsaturated solutions contain less than this amount of solvent.

Supersaturated solutions contain more than this amount of solute. A supersaturated solution can be produced by dissolving at elevated temperatures and slowly allowing the solution to cool to room temperature. Such solutions tend to be very unstable. The excess solute will suddenly "fall out" of solution if the solution is agitated or a few seed particles of solute are added. The term *precipitation* is used to describe an event in which dissolved material falls out of solution. Usually this material settles to the bottom of the container.

Methods of Indicating Concentration

A variety of methods are used to designate the concentration of solutes in solutions. These methods use different units and different relationships between solute and solvent, or between solute and solution.

Some of the more familiar terms used to indicate concentration are mainly descriptive, and are of little use quantitatively. These terms include concentrated, dilute, weak and strong. **Dilute** solutions contain small amounts of solute; **concentrated** solutions contain large amounts of solute. **Weak** solutions are dilute; **strong** solutions are concentrated.

Molarity. The **molarity** (M) of a solution is one of several terms used to indicate the amount of solute in a given amount of solution. A one molar solution (1. 00M) contains one mole of *solute* dissolved in enough solvent to make 1.00 liter of *solution.* A general expression for molarity is:

$$M = \frac{\text{number of moles of solute}}{\text{liters of solution}}$$

A two molar (2M) solution contains 2 moles of solute per liter of solution. A 0.1 molar (0.1M) solution contains 0.1 mole per liter of solution.

SAMPLE PROBLEM

What is the molarity of a solution of sodium oxalate, $Na_2C_2O_4$, containing 33.5 grams of solute in 100.0 mL of solution?

Solution:

1. Calculate the mass of solute in one liter of solution:

$$\frac{33.5 \text{ g}}{100\text{mL}} \times 1000 \text{ mL/liter} = 335 \text{ g/liter}$$

2. Calculate the mass of one mole of $Na_2C_2O_4$:

mole mass $= (2 \times 23) + (2 \times 12) + (4 \times 16) = 134$ g

3. Calculate the number of moles of solute per liter of solution (definition of molarity):

$$\frac{335 \text{ g}}{1 \text{ liter}} \times \frac{1 \text{ mole}}{134 \text{ g}} = 2.50 \text{ moles/liter} = \textbf{2.50 M}$$

If the molarity of a solution is known, the number of moles in a given volume of that solution can be calculated as follows:

moles of solute = molarity × volume in liters

SAMPLE PROBLEM

How many moles of NaOH are contained in 200mL of a 0.1M solution of NaOH?

Solution:

1. Find the volume of the solution in liters:

$$200 \text{ mL} \times \frac{1 \text{ liter}}{1000 \text{ mL}} = 0.2 \text{ liter}$$

2. Substitute this volume into the equation:

moles of solute = 0.1 moles/liter × 0.2 liter
$$= \textbf{.02 moles NaOH}$$

Molal Solutions. A one **molal** (1.00 m) solution contains one mole of solute dissolved in 1000 grams of *solvent*. This method of expressing concentration is used in studies dealing with the effect of dissolved particles on certain properties of the solvent (colligative properties).

SAMPLE PROBLEM

What is the molality of a solution of glucose, $C_6H_{12}O_6$, that contains 0.09 gram of solute dissolved in 5.0 grams of water?

Solution:

1. Calculate the mass of solute in 1000 g of solvent:

$$\frac{0.09° \text{ g of solute}}{5.0 \text{ g of solvent}} \times 1000 = 18 \text{ g solute per 1000 g solvent}$$

2. Calculate the mole mass of solute:

mole mass $= (6 \times 12) + (12 \times 1) + (6 \times 16) = 180$ g/mole

3. Calculate the number of moles of solute per 1000 grams of solvent (definition of molality):

$$\frac{18 \text{ g solute}}{1000 \text{ g solvent}} \times \frac{1 \text{ mole solute}}{180. \text{ g solute}} = .10 \text{ m}$$

Effect of Solute on Solvent

Certain properties of solvents are affected by the presence of dissolved particles. Those properties that depend on the relative *number* of particles rather than the nature of the particles are called **colligative properties.** Colligative properties include boiling point, freezing point, vapor pressure, and osmotic pressure.

Boiling point elevation. The presence of a nonvolatile solute raises the boiling point of the solvent. (Non-volatile substances are those that do not evaporate readily at room temperature and pressure.) One mole of nonvolatile particles in 1000 grams of water (1.00 molal solution) will raise the boiling point of the water by .5l2C°. This is sometimes referred to as the boiling point elevation constant for water.

Freezing point depression. The freezing point of a solvent is lowered, or depressed, in proportion to the number of particles dissolved in it. One thousand grams of water has its freezing point lowered 1.86C° by the presence of one mole of solute particles.

Abnormal behavior of electrolytes. In solution, ionic solutes seem to produce larger changes in the colligative properties of their solvents than do nonionic, or molecular, solutes. This is due to the fact that ionic substances dissociate, or break apart, when they enter solution, thereby adding more particles to the solution.

QUESTIONS

1. How many grams of ammonium chloride (gram formula mass = 53.5 g) are contained in 0.500 L of a 2.00 M solution?

   (1) 10.0 g   (2) 26.5 g   (3) 53.5 g   (4) 107 g

2. What is the molarity of an H_2SO_4 solution if 0.25 liter of the solution contains 0.75 mole of H_2SO_4?

   (1) 0.33 M   (2) 0.75 M   (3) 3.0 M   (4) 6.0 M

3. What is the total number of grams of NaOH (formula mass = 40.) needed to make 1.0 liter of a 0.20 M solution?

   (1) 20. g   (2) 2.0 g   (3) 80. g   (4) 8.0 g

4. As additional $KNO_3(s)$ is added to a saturated solution of KNO_3 at constant temperature, the concentration of the solution

   (1) decreases   (2) increases   (3) remains the same

5. What is the total number of grams of KCl (formula mass = 74.6) in 1.00 liter of 0.200 molar solution?

   (1) 7.46 g   (2) 14.9 g   (3) 22.4 g   (4) 29.8 g

6. If 0.50 liter of a 12-molar solution is diluted to 1.0 liter, the molarity of the new solution is   (1) 2.4   (2) 6.0   (3) 12   (4) 24

7. A 1 kilogram sample of water will have the highest freezing point
    when it contains

    (1) 1×10^{17} dissolved particles   (2) 1×10^{19} dissolved particles
    (3) 1×10^{21} dissolved particles   (4) 1×10^{23} dissolved particles

8. Compared to the normal freezing point and boiling point of water,
    a 1-molal solution of sugar in water will have a

    (1) higher freezing point and a lower boiling point
    (2) higher freezing point and a higher boiling point
    (3) lower freezing point and a lower boiling point
    (4) lower freezing point and a higher boiling point

9. Which ratio of solute-to-solvent could be used to prepare a
    solution with the highest boiling point?

    (1) 1 g of NaCl dissolved per 100 g of water
    (2) 1 g of NaCl dissolved per 1000 g of water
    (3) 1 g of $C_{12}H_{22}O_{11}$ dissolved per 100 g of water
    (4) 1 g of $C_{12}H_{22}O_{11}$ dissolved per 1000 g of water

10. Which expression defines the molality (m) of a solution?

    (1) $\dfrac{\text{grams of solute}}{\text{kg of solution}}$ (2) $\dfrac{\text{moles of solute}}{\text{kg of solution}}$

    (3) $\dfrac{\text{grams of solute}}{\text{kg of solvent}}$ (4) $\dfrac{\text{moles of solute}}{\text{kg of solvent}}$

11. A student dissolves 1.0 mole of sucrose ($C_{12}H_{22}O_{11}$) in 1,000
    grams of water at 1.0 atmosphere. Compared to the boiling point
    of pure water, the boiling point of the resulting solution is

    (1) 0.52 C° lower   (2) 1.86 C° lower   (3) 0.52 C° higher
    (4) 1.86 C° higher

12. Which solute, when added to 1,000 grams of water, will produce a solution with the highest boiling point?
(1) 29 g of NaCl   (2) 58 g of NaCl
(3) 31 g of $C_2H_6O_2$ (4) 62 g of $C_2H_6O_2$

13. What occurs as a salt dissolves in water?
(1) The number of ions in the solution decreases, and the freezing point decreases.
(2) The number of ions in the solution decreases, and the freezing point increases.
(3) The number of ions in the solution increases, and the freezing point decreases.
(4) The number of ions in the solution increases, and the freezing point increases.

UNIT 6. KINETICS AND EQUILIBRIUM

KINETICS

Chemical kinetics deals with:

- The rates of chemical reactions
- The mechanisms by which the reactions occur

A **reaction rate** depends on several variables, including the nature of the reactants, the concentration of the reactants, and the temperature of the system. Rate of reaction is the change in concentration of a given substance (reactant or product) per unit time. Reaction rate is measured in terms of moles of reactant consumed (or product formed) per liter per second.

Very few chemical reactions occur as directly as an equation for that reaction indicates. In many cases, an equation represents a *net* reaction, which is a summation of several intermediate reactions. These intermediate equations represent the actual *reaction mechanism* of the overall reaction.

Role of Energy in Reactions

In order for a chemical reaction to begin, energy is needed. As a result of the reaction, energy may be released or absorbed by the system. A complete picture of the energy factors in a chemical reaction is best illustrated by a potential energy diagram in which the potential energy of the substances involved in the reaction is plotted against a time sequence.

Activation energy. Activation energy is the minimum energy needed to cause a reaction to begin.

Heat of reaction. Heat of reaction is the difference between the potential energy of the products and that of the reactants. This potential energy is also known as **enthalpy**. It was once referred to as "heat content." This obsolete and misleading term accounts for the use of the symbol H in the equation:

$$\Delta H = H_{products} - H_{reactants}$$

The Greek letter Δ, called *delta,* denotes a "difference." ΔH is the symbol for Heat of Reaction. A number of ΔH's for reactions are listed on table G and I of the Reference Tables.

Exothermic reactions are those that release energy. In such reactions, the potential energy of the products is lower than that of the reactants. In exothermic reactions, the sign of ΔH is negative. Consider the reaction:

$$\frac{1}{2}N_2(g) + O_2(g) \rightarrow NO_2(g) + 7.9 \text{ kcal}$$

The reaction releases heat. It is exothermic. ΔH for this reaction is -7.9 kcal/mole NO_2.

Potential Energy Diagram For An Exothermic Reaction

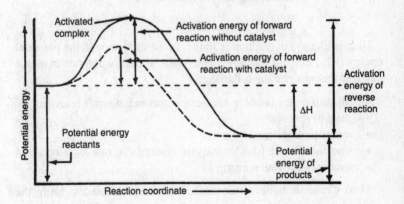

Endothermic reactions are those that absorb energy. In such reactions, the potential energy of the products is greater than that of the reactants. In endothermic reactions, the sign of ΔH is positive. Consider the reaction:

$$\tfrac{1}{2}H_2(g) + \tfrac{1}{2}I_2(g) + 6.30 \text{ kcal} \rightarrow HI(g)$$

This reaction is endothermic, absorbing energy.

ΔH is + 6.30 kcal mole HI

Potential Energy Diagram For An Endothermic Reaction

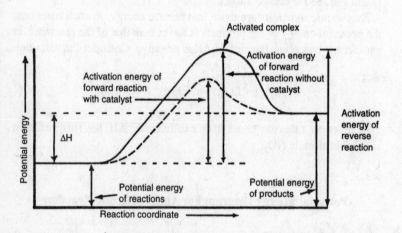

Heat (enthalpy) of reaction is shown as the difference in the potential energy (P.E.) in these reactions. The potential energy diagram makes it easier to demonstrate the following:

- the difference between exothermic and endothermic reactions.
- heat of reaction
- activation energy
- the manner in which catalysts change the rate of a reaction (reduces activation energy)

Most Chemists believe that activation energy is used to bring the reactants together to form an intermediate substance of greater energy. The **activated complex** is extremely unstable and immediately decomposes to form the products of the reaction.

Factors Affecting Rates of Reaction

It is believed that in order for particles (atoms, molecules, or ions) to react, they must collide with each other. Calculations have shown, however, that when the number of collisions is increased, the rate of reaction is *not* increased proportionately. This discovery has led to the conclusion that not all collisions are *effective* in bringing about reaction. In order to be effective, collisions must have (1) enough energy and (2) the proper orientation of the colliding particles.

The effectiveness of collisions is determined by several factors:

Nature of the reactants. Chemical reactions occur by the breaking and rearranging of existing bonds and by the forming of new bonds. The fewer the rearrangements that occur, the faster the reaction takes place. Reactions involving ions in aqueous solution are usually extremely rapid, probably because there are no bonds to be broken. Reactions between molecules require the breaking of bonds. At room temperature, reactions such as the one between hydrogen (H_2) and oxygen (O_2) are usually very slow.

Concentration. An increase in the concentration of a reactant increases the frequency of collisions. The result is an increase in reaction rate. It is customary in chemical kinetics to express concentrations in moles per liter (molarity). Brackets around a formula or symbol are used to indicate concentration. $[Cl^-]$ means the concentration of the chloride ion in molarity.

In reactions involving gases in closed systems, *increased pressure* results in increased concentration. The net result is an increase in the rate of reaction.

Temperature. Increasing the temperature of the system always increases the rate of reaction. Since temperature represents the average kinetic energy of the particles, the effectiveness of the collisions, as well as the number of collisions is increased.

Surface Area. As one might expect, increasing the surface area of reactants provides greater opportunity for collisions to occur. This can easily be demonstrated in the reaction between zinc metal and hydrochloric acid to produce hydrogen gas. Reducing the size of the zinc particles increases the surface area of the zinc. Evidence of the increase in reaction rate is provided by the increased bubbling resulting from hydrogen formation.

Catalysts. Catalysts, by definition, are substances that increase the rate of reaction. This increased rate is brought about through a change in the reaction mechanism so that less activation energy is required. Catalyzed reactions take place more rapidly than uncatalyzed reactions because of the reduction in activation energy. The potential energy of the reactants and the potential energy of the products remain the same. Therefore, the heat of reaction is the same, whether catalyzed or not.

QUESTIONS

1. An increase in the temperature increases the rate of a chemical reaction because the collisions in this reaction increase in (1) number only (2) effectiveness, only (3) both number and effectiveness (4) neither number nor effectiveness
2. The vapor pressure of a liquid at a given temperature is measured when the rate of evaporation of the liquid becomes (1) less than the rate of condensation (2) greater than the rate of condensation (3) equal to the rate of condensation (4) equal to a zero rate of condensation
3. In a chemical reaction, the difference between the potential energy of the products and the potential energy of the reactants is the (1) heat of reaction (2) heat of fusion (3) free energy (4) activation energy

   Base your answers to questions 4 and 5 on the potential energy diagram below

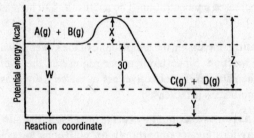

4. The potential energy of the activated complex is equal to the sum of (1) X + Y (2) X + W (3) X + Y + W (4) X + W + Z
5. The reaction A(g) + B(g) → C(g) + D(g) + 30 kcal has a forward activation energy of 20 kcal. What is the activation energy for the reverse reaction? (1) 10 kcal (2) 20 kcal (3) 30 kcal (4) 50 kcal

6. The graph below represents a chemical reaction.

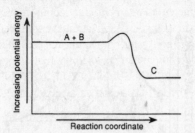

This reaction is best described as (1) endothermic, because energy is absorbed (2) endothermic, because energy is released (3) exothermic, because energy is absorbed (4) exothermic, because energy is released

7. The potential energy diagram shown below represents the reaction $R + S + \text{energy} \rightarrow T$.

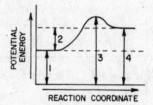

Which numbered interval represents the potential energy of the product T? (1) 1 (2) 2 (3) 3 (4) 4

8. A potential energy diagram is shown below.

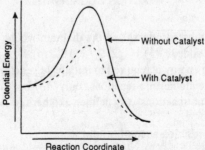

Which reaction would have the *lowest* activation energy? (1) the forward catalyzed reaction (2) the forward uncatalyzed reaction (3) the reverse catalyzed reaction (4) the reverse uncatalyzed reaction

9. The potential energy diagram of a chemical reaction is shown below.

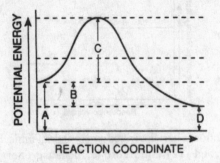

Which letter in the diagram represents the heat of reaction (ΔH)? (1) A (2) B (3) C (4) D

10. The potential energy diagram below represents the reaction $2KClO_3 \rightarrow 2KCl + 3O_2$

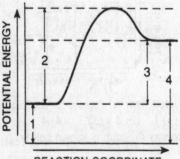

Which numbered interval on the diagram would change when a catalyst is added? (1) 1 (2) 2 (3) 3 (4) 4

11. Activation energy is required to initiate (1) exothermic reactions, only (2) endothermic reactions, only (3) both exothermic and endothermic reactions (4) neither exothermic nor endothermic reactions

12. Given the reaction at equilibrium:

$$CO_3^{2-}(aq) + H_2O(\ell) \rightleftarrows HCO_3^-(aq) + OH^-(aq)$$

Which statement is always true? (1) $[CO_3^{2-}]$ is less than $[OH^-]$. (2) $[CO_3^{2-}]$ is less than $[HCO_3^-]$. (3) The rate of the forward reaction is less than the rate of the reverse reaction. (4) The rate of the forward reaction equals the rate of the reverse reaction.

13. Given the reaction: $A + B \rightarrow C + D$
The reaction will most likely occur at the greatest rate if A and B represent (1) nonpolar molecular compounds in the solid phase (2) ionic compounds in the solid phase (3) solutions of nonpolar molecular compounds (4) solutions of ionic compounds

14. According to Reference Table G, which compound forms spontaneously from its elements? (1) $C_2H_2(g)$ (2) $C_2H_4(g)$ (3) ICl(g) (4) NO(g)

15. What is the heat of formation of $H_2O(\ell)$, in kilocalories per mole, at 1 atmosphere and 298 K? (1) -79.7 (2) -68.3 (3) -56.7 (4) -54.6

16. According to Reference Table I, the dissolving of which salt is accompanied by the release of energy? (1) LiBr (2) NH_4Cl (3) NaCl (4) KNO_3

EQUILIBRIUM

Since most reactions proceed in both directions, it is necessary to study the rates of both the forward and reverse reactions. When the forward and reverse rates are equal in such *reversible systems,* the system is said to be in a state of **dynamic equilibrium**. The word dynamic suggests motion. Dynamic equilibrium refers to a state of balance between two opposing activities in which the concentration of both reactants and products remain *constant*. These concentrations are not necessarily equal, however. A state of equilibrium may exist in which quantities of reactants and products are quite different. Furthermore, the reversibility of reactions permits the attainment of equilibrium from either the forward or the reverse reaction.

Phase Equilibrium

When substances change phase, the changes are reversible. In closed systems, a state of equilibrium between phases may be achieved. Whenever liquids or solids are confined in a closed container, the condition will be reached in which there are enough particles in the gas (vapor) phase to cause the rate of return to the original phase to be equal to the rate of escape. This condition will result in a vapor pressure that is temperature dependent and characteristic for the particular solid or liquid involved.

Solution Equilibrium

Gases in liquids. Liquids containing dissolved gases in closed systems will reach equilibrium not only between the liquid and its gas phase but also between the dissolved gas and the undissolved gas above the liquid. This equilibrium is affected by temperature and pressure, since increased temperature and/or reduced pressure reduces the solubility of gases in liquids. Low temperatures and high pressures favor solution of gases in liquids. Carbonated beverages, therefore, maintain their carbonation best when tightly covered and chilled.

Solids in liquids. When a solution of a solid in a liquid contains all the solid that will dissolve at existing conditions, it is said to be a saturated solution. If additional solute is added, it will fall to the bottom of the container. A condition of equilibrium may be reached between the dissolved and undissolved solute. In this condition the processes of dissolving and crystallizing of the solute occur at equal rates.

Solubility. The concentration of the solute in the saturated solution described above is referred to as the **solubility** of the solute in that liquid. To review the vocabulary of solutions:

- The solute is the substance dissolved.
- The solvent is the substance in which the solute is dissolved.
- The solution is the homogeneous mixture that results.

Chemical Equilibrium

In any chemical reaction, equilibrium is recognized when observable changes, such as color, temperature, pressure etc. (known as macroscopic changes) no longer occur. At this point the forward and reverse reactions are occurring at equal rates.

Le Chatelier's Principle. Chemists refer to changes in concentration, pressure, or temperature as **applied stresses**. Le Chatelier's Principle is a generalization that describes what happens to a system subjected to such stresses. When a stress is applied to a system in equilibrium, the reaction will try to shift in the direction that will relieve the stress. Equilibrium will then be re-established at a different point.

This principle, and how it operates, is well demonstrated in the *Haber process* for the manufacture of ammonia (NH_3) from nitrogen (N_2) and hydrogen (H_2). The equation for the reaction is:

$$N_2 (g) + 3H_2 (g) \rightleftarrows 2NH_3 (g) + 22 \text{ kcal}$$

The equation for the reaction provides the following information:

1. The reaction is *exothermic,* since there is a release of 11 kcal of heat for each mole of NH_3 produced. If the reaction followed the rules of stoichiometry, 4 moles of reactant would yield 2 moles of product, resulting in a reduced volume.

2. Since the reactants and the products are gases, they are subject to both pressure and temperature changes.

    a. *Effect of concentration.* Increasing the concentration of one substance in a reaction at equilibrium will cause the reaction to proceed in the direction that will use up the increase. In time, a new equilibrium will be established. In the Haber process, increasing the concentration of nitrogen or hydrogen will increase the rate of ammonia production. Of course, if the system remains closed, the increased concentration of ammonia will cause an increase in the reverse reaction until a new equilibrium is reached. If ammonia is removed, however, the resulting decrease in concentration of product will further the forward reaction so as to increase the output of ammonia.

        Products may be removed from a reaction by (1) the formation of a gas (2) the formation of an insoluble product (precipitate) by the reaction of dissolved reactants, or (3) the formations of a non-electrolyte such as water. Such reactions are called reactions that go to completion.

    b. *Effect of pressure.* Changes in pressure affect only the gaseous components in an equilibrium system. Increased pressure favors the reaction that results in a smaller volume caused by fewer number of gaseous molecules. Increasing the pressure in the Haber process, therefore, will favor the forward reaction, since the production of ammonia results in reduced volume. This reduction in volume relieves the stress caused by the increased pressure. In reactions in which the volume of reactant(s) is equal to the volume of product(s), changes in pressure have no effect and therefore the equilibrium will not shift in either direction.

    c. *Effect of temperature.* When the temperature of a system in equilibrium is raised, the equilibrium shifts in the direction that will absorb that heat. An endothermic reaction is favored. A decrease in temperature of a system in equilibrium favors an exothermic reaction. For the Haber process, an increase in temperature causes the reaction to shift to the reactants favoring the decomposition of ammonia.

    d. *Effect of catalysts.* The addition of a catalyst to a system may cause equilibrium to be reached more rapidly. There will be *no net change* in the equilibrium concentration, because catalysts increase rates of forward and reverse reactions equally.

QUESTIONS

1. A solution in which equilibrium exists between undissolved and dissolved solute is always
   (1) saturated   (2) unsaturated   (3) dilute   (4) concentrated

2. A liquid in a stoppered flask is allowed to stand at constant temperature until the liquid level in the flask remains constant. Which condition then exists in the flask? (1) Only liquid is evaporating. (2) Only vapor is condensing (3) The rate of condensation is greater than the rate of evaporation. (4) The rate of condensation is equal to the rate of evaporation.

3. A flask at 25°C is partially filled with water and stoppered. After a period of time the water level remained constant. Which relationship best explains this observation?
   (1) The rate of condensation exceeds the rate of evaporation.
   (2) The rates of condensation and evaporation are both zero.
   (3) The rate of evaporation exceeds the rate of condensation.
   (4) The rate of evaporation equals the rate of condensation.

4. Given the equation:

   $$H_2(g) + I_2(g) \rightleftarrows 2HI(g)$$

   Which statement is always true when this reaction has reached chemical equilibrium? (1) $[H_2] \times [I_2] > [HI]$.
   (2) $[H_2] \times [I_2] < [HI]$. (3) $[H_2]$, $[I_2]$, and $[HI]$ are all equal.
   (4) $[H_2]$, $[I_2]$, and $[HI]$ remain constant.

5. Given the reaction:

   $$HC_2H_3O_2(aq) + H_2O \rightleftarrows H_3O^+(aq) + C_2H_3O_2^-(aq)$$

   When the reaction reaches a state of equilibrium, the concentrations of the reactants (1) are less than the concentrations of the products (2) are equal to the concentrations of the products (3) begin decreasing (4) become constant

6. Which change would most likely increase the rate of a chemical reaction? (1) decreasing a reactant's concentration (2) decreasing a reactant's surface area (3) cooling the reaction mixture (4) adding a catalyst to the reaction mixture

7. Given the system $CO_2(s) \rightleftarrows CO_2(g)$ at equilibrium. As the pressure increases at constant temperature, the amount of $CO_2(g)$ will (1) decrease (2) increase (3) remain the same

8. Which system at equilibrium will shift to the right when the pressure is increased?

(1) $NaCl(s) \overset{H_2O}{\rightleftarrows} Na^+ (aq)\, Cl^- (aq)$

(2) $C_2H_5OH\, (\ell) \overset{H_2O}{\rightleftarrows} C_2H_5OH\, (aq)$

(3) $NH_3(g) \overset{H_2O}{\rightleftarrows} NH_3\, (aq)$

(4) $C_6H_{12}O_6\, (s) \overset{H_2O}{\rightleftarrows} C_6H_{12}O_6\, (aq)$

9. Given the reaction at equilibrium:

$$2SO_2(g) + O_2(g) \rightleftarrows 2SO_3(g) + heat$$

The rate of the forward reaction can be increased by adding more SO_2 because the (1) temperature will increase (2) number of molecular collisions between reactants will increase (3) reaction will shift to the left (4) forward reaction is endothermic

10. Given the reaction at equilibrium:

$$CO(g) + \tfrac{1}{2}O_2(g) \rightleftarrows CO_2(g) + 67.7\, kcal$$

As the temperature increases, the rate of the forward reaction (1) decreases (2) increases (3) remains the same

11. The diagram below shows a bottle containing $NH_3(g)$ dissolved in water. How can the equilibrium $NH_3(g) \rightleftarrows NH_3(aq)$ be reached?

(1) Add more water (2) Add more $NH_3(g)$ (3) Cool the contents (4) Stopper the bottle

12. Given the solution at equilibrium:

$$BaSO_4(s) \rightleftarrows Ba^{2+} (aq) + SO_4^{2-} (aq)$$

The K_{sp} of $BaSO_4(s)$ at 25°C is 1.08×10^{-10}. Which change will affect the K_{sp} value? (1) increasing the $[Ba^{2+}]$ (2) increasing the $[SO_4^{2-}]$ (3) increasing the pressure (4) increasing the temperature

13. What effect does the addition of a catalyst have on a chemical reaction at equilibrium? (1) It increases the rate of the forward reaction, only (2) It increases the rate of the reverse reaction, only (3) It increases the rate of both the forward and reverse reactions. (4) It decreases the rate of both the forward and reverse reactions.

14. At which temperature can water contain the most dissolved oxygen at a pressure of 1 atomsphere? (1) 10.°C (2) 20.°C (3) 30.°C (4) 40.°C

15. The reaction:
$Ba(NO_3)_2(aq) + Na_2SO_4(aq) \rightarrow 2NaNO_3(aq) + BaSO_4(s)$ goes to completion because a (1) gas is formed (2) precipitate is formed (3) nonionized product is formed (4) soluble salt is formed

16. Consider the following reaction:

$$A (g) + 2B (g) \rightleftarrows C (g) + heat$$

The reaction is carried out in a high-pressure container at constant temperature. The initial concentration of A is 1.0 mole/liter. At equilibrium, the concentration of C is .25 mole/liter.

The concentration of A at equilibrium in moles per liter is
(1) 1.0   (2) 0.75   (3) 0.50   (4) 0.25

17. According to Reference Table D, a temperature change from 10°C to 30°C would have the *least* effect on the solubility of
(1) NaCl   (2) $KClO_3$ (3) NH_3 (4) SO_2

18. According to Reference Table D, which is the best description of the system prepared by dissolving 30 grams of $NH_3(g)$ in 100 grams of water at 20°C? (1) a saturated solution of NH_3 with no excess $NH_3(g)$ (2) a saturated solution of NH_3 in contact with excess $NH_3(g)$ (3) an unsaturated solution of NH_3 with no excess $NH_3(g)$ (4) an unsaturated solution of NH_3 in contact with excess $NH_3(g)$

Law of Chemical Equilibrium. With respect to reversible reactions at equilibrium, as long as the temperature remains constant, there is a mathematical relationship that describes the system. This relationship is known as the Law of Chemical Equilibrium.

Suppose that substance A reacts with substance B to produce substances C and D according to the equation

$$aA + bB \quad \rightleftarrows \quad cC + dD$$

in which a, b, c, and d are coefficients in the equation. As such, they represent the number of moles of each substance in the reaction.

To write the equilibrium expression for this reaction:

1. Write the symbols for the products in the numerator and the symbols for the reactants in the denominator.
2. Place square brackets around each symbol. These brackets indicate concentration in moles/liter (molarity).
3. Raise the concentration of each substance to the power of its coefficient.
4. Set the expression equal to K_{eq}. K_{eq} is the equilibrium constant for the reaction at a constant temperature.

$$K_{eq} = \frac{[C]^c[D]^d}{[A]^a[B]^b}$$

Apply these rules to the equation for the reaction between hydrogen gas and iodine gas to produce hydrogen iodide gas:

$$H_2(g) + I_2(g) \quad \rightleftarrows \quad 2HI(g) \qquad\qquad K_{eq} = \frac{[HI]^2}{[H_2][I_2]}$$

Usually, equilibrium constants are written with subscripts that indicate the kind of reaction. For example:

K_a for the ionization of acids
K_b for the ionization of bases
K_w for the ionization of water
K_{eq} for chemical reactions

Concentrations of solids and liquids in reactions are always constant. Thus, they do not appear in the equilibrium expression for the reaction. Only concentrations of gases and aqueous solutions appear in K_{eq} expression.

Consider the reaction

$$4H_2O(g) + 3Fe(s) \quad \rightleftarrows \quad 4H_2(g) + Fe_3O_4(s)$$

The equilibrium expression for this reaction is written

$$K_{eq} = \frac{[H_2]^4}{[H_2O]^4}$$

Generalizations on the Law of Chemical Equilibrium:

1. When the value of K_{eq} is large, the numerator is much larger than the denominator. Therefore, the concentration of products must be greater than the concentration of reactants. It is said that the products are favored.

2. When the value of K_{eq} is small, the reactants are favored.

3. If temperature remains constant, changes in concentration of any of the substances involved do not change the value of K_{eq}. Instead, the concentration of the other substances would change to compensate.

4. Temperature changes *do* cause a change in K_{eq} because such changes affect the rates of the opposing reactions in different ways.

Spontaneous Reactions

Whether reactions proceed or not, seems to depend on the balance between two basic natural tendencies:

1. the drive toward greater stability (reduced potential energy)

2. the drive toward less organization (increased entropy)

Energy changes. Systems in conditions of great energy tend to change to conditions of low energy. Given ΔH as the symbol for energy change in chemical systems, this tendency favors reactions in which ΔH is negative; that is, *exothermic* reactions.

Entropy changes. Systems tend to change from conditions of great order to conditions of low order. This tendency is interpreted as a change from low entropy to high entropy. The term *randomness* is frequently used to describe entropy. The more random the system, the higher the entropy. High entropy is favored by increased temperature.

The symbol for entropy is S. A positive ΔS indicates a phase change of increasing randomness:

$$\text{solid} \text{-----}> \text{liquid} \text{-----}> \text{gas} \qquad \Delta S = (+)$$

At constant temperature, substances in a system tend to change phase so that, in its final state, the system has higher randomness (entropy) than it had in its initial state.

Solubility Product Constant (K_{sp})

Solubility product constant is a special case of the equilibrium constant used as a measure of the concentration of slightly soluble salts in water. The equation for the equilibrium reaction between the excess solid and its ions in the case of $PbCl_2$ is derived from the equation:

$$PbCl_2(s) \rightleftarrows Pb^{2+}(aq) + 2Cl^-(aq)$$

The solubility product constant is

$$K_{sp} = [Pb^{2+}][Cl^-]^2$$

The $PbCl_2$ does not appear in the expression because it is a solid.

The value K_{sp} is used to compare solubilities. The lower the value of K_{sp}, the lower the solubility. Given the following values:

$$K_{sp} \text{ for } CaSO_4 = 3 \times 10^{-5}$$
$$K_{sp} \text{ for } BaSO_4 = 1 \times 10^{-10}$$

the $BaSO_4$ is less soluble than $CaSO_4$ at the same temperature.

A variety of equilibrium constants are listed on Table M of the Reference Tables.

QUESTIONS

1. What is the correct equilibrium constant expression for the reaction below?

$$N_2(g) + 3H_2(g) \rightleftarrows 2NH_3(g)$$

(1) $K_{eq} = \dfrac{[N_2][H_2]^3}{[NH_3]^2}$ 　　　(2) $K_{eq} = \dfrac{[NH_3]^2}{[N_2][H_2]^3}$

(3) $K_{eq} = \dfrac{[N_2][3H_2]}{[2NH_3]}$ 　　　(4) $K_{eq} = \dfrac{[2NH_3]}{[N_2][3H_2]}$

2. Which is the correct equilibrium expression for the reaction

$$3A(g) + 4B(g) \rightleftarrows 2C(g) + 5D(g) ?$$

(1) $K_{eq} = \dfrac{[A]^3[B]^4}{[C]^2[D]^5}$　　　　(2) $K_{eq} = \dfrac{[A]^3[C]^2}{[B]^4[D]^5}$

(3) $K_{eq} = \dfrac{[C]^2[D]^5}{[A]^3[B]^4}$　　　　(4) $K_{eq} = \dfrac{[B]^4[D]^5}{[A]^3[C]^2}$

3. Given the system at chemical equilibrium:

$$2O_3(g) \rightleftarrows 3O_2(g) \ (K_{eq} = 2.5 \times 10^{12})$$

The concentration of O_3 and O_2 must be (1) constant (2) equal
(3) increasing (4) decreasing

4. Given the reaction at equilibrium:

$$CH_3COOH(aq) + H_2O(\ell) \rightleftarrows H_3O^+(aq) + CH_3COO^-(aq)$$

The addition of which ion will cause an increase in the concentration of CH_3COO^- (aq)? (1) H_3O^+ (2) Cl^- (3) Na^+ (4) OH^-

5. Which of the following compounds has the smallest K_{sp} value?
(1) AgCl (2) AgBr (3) $PbCl_2$ (4) PbI_2

6. Based on Reference Table M, which of the following is the *least* soluble compound? (1) AgBr (2) AgCl (3) Ag_2CrO_4 (4) AgI

7. Which equilibrium constant indicates an equilibrium mixture that favors the formation of products? (1) $K_{eq} = 1 \times 10^{-5}$
(2) $K_{eq} = I \times 10^{-1}$ (3) $K_{sp} = 1 \times 10^{0}$ (4) $K_{sp} = 1 \times 10^{5}$

8. Given the reaction at equilibrium:

$$H_2(g) + \tfrac{1}{2}O_2(g) \rightleftarrows H_2O(g) + heat$$

The value of the equilibrium constant for this reaction can be changed by
(1) changing the pressure (2) changing the temperature
(3) adding more O_2 (4) adding a catalyst

9. One mole of each of the salts below is added to a liter of water. Which salt will produce the highest concentration of carbonate ions (CO_3^{2-})?
(1) $MgCO_3, K_{sp} = 1.2 \times 10^{-5}$ (2) $CaCO_3, K_{sp} = 5.0 \times 10^{-9}$
(3) $BaCO_3, K_{sp} = 2.6 \times 10^{-9}$ (4) $ZnCO_3, K_{sp} = 1.4 \times 10^{-11}$

10. Which reaction has the equilibrium expression

$$K = \frac{[A][B]^2}{[AB_2]}$$

(1) $AB_2(g) \rightleftarrows A(g) + 2B(g)$ (2) $2AB(g) \rightleftarrows A(g) + B_2(g)$

(3) $A(g) + 2B(g) \rightleftarrows AB_2(g)$ (4) $A(g) + B_2(g) \rightleftarrows 2AB(g)$

11. Given the reaction: $A + B \rightleftarrows AB$

The greatest amount of AB would be produced if the equilibrium constant of the reaction is equal to

(1) 1.0×10^{-5} (2) 1.0×10^{-1} (3) 1.0×10^{1} (4) 1.0×10^{5}

12. Given the reaction:

$$HNO_2(aq) \rightleftarrows H^+(aq) + NO_2^-(aq)$$

The ionization constant, K_a, is equal to

(1) $\dfrac{[HNO_2]}{[H^+][NO_2^-]}$ (2) $\dfrac{[H^+][NO_2^-]}{[HNO_2]}$

(3) $\dfrac{[NO_2^-]}{[H^+][HNO_2]}$ (4) $\dfrac{[H^+][HNO_2]}{[NO_2^-]}$

13. Which compound has the smallest K_{sp} at 298 K? (1) AgCl (2) AgI   (3) $PbCl_2$ (4) PbI_2

14. Given the reaction at equilibrium:

$$A(g) + B(g) + heat \rightleftarrows AB(g)$$

As the pressure increases at constant temperature, the value of the equilibrium constant
(1) decreases   (2) increases   (3) remains the same

15. As a catalyst is added to a system at equilibrium, the value of the equilibrium constant
(1) decreases   (2) increases   (3) remains the same

16. What is the equilibrium expression for the reaction

$4Al(s) + 3O_2(g) \rightleftharpoons 2Al_2O_3(s)$?

(1) $K_{eq} = [O_2]^3$ (2) $K_{eq} = [3O_2]$

(3) $K_{eq} = \dfrac{1}{[O_2]^3}$ (4) $K_{eq} = \dfrac{1}{[3O_2]}$

17. Which change represents an increase of entropy?
(1) $I_2(s) \rightarrow I_2(g)$ (2) $I_2(g) \rightarrow I_2(\ell)$ (3) $H_2O(g) \rightarrow H_2O(\ell)$
(4) $H_2O(\ell) \rightarrow H_2O(s)$

18. Given the reaction at equilibrium:

$$AgI(s) \rightleftharpoons Ag^+(aq) + I^-(aq)$$

Which is the correct solubility product constant expression for the reaction?

(1) $K_{sp} = [Ag^+][I^-]$ (2) $K_{sp} = \dfrac{1}{[Ag^+][I^-]}$

(3) $K_{sp} = [Ag^+] + [I^-]$ (4) $K_{sp} = \dfrac{1}{[Ag^+] + [I^-]}$

19. The reaction $CH_3COOH(aq) \rightleftharpoons CH_3COO^-(aq) + H^+(aq)$ has a K_a equal to 1.8×10^{-5} at $25°$ C. In a solution of this acid at $25°$ C, the concentration of CH_3COOH is (1) less than the concentration of H^+ ions (2) equal to the concentration of H^+ ions (3) greater than the concentration of CH_3COO^- ions (4) equal to the concentration of CH_3COO^- ions

ADDITIONAL MATERIALS

Free Energy Change

Free energy change is the difference between the energy change (ΔH) and the product of the absolute temperature and the entropy change (ΔS). The symbol ΔG has been assigned to this change, where G represents the *free energy*. Therefore, at a constant temperature, T:

$$\Delta G = \Delta H - T\Delta S$$

Predicting Spontaneous Reactions

In order for a change in a system to occur spontaneously, ΔG must be negative. The value of ΔG is zero for systems at equilibrium.

When the two factors, the drive toward lower energy and the drive toward greater entropy, cannot be satisfied at the same time, the change that will be favored will depend on the temperature of the system.
For example:

1. Considering only enthalpy (ΔH) freezing of water is exothermic and therefore should occur at any temperature. However, the drive toward higher entropy favors melting. At temperatures below $0°$ C, the *energy* change dominates and freezing *does* occur. At temperatures above $0°$ C, the *entropy* change dominates and the ice will melt.

2. The decomposition of $KClO_{3(s)}$ to $KCl_{(s)}$ and $O_{2(g)}$ is endothermic and the energy change would oppose the reaction. However, the formation of a gas (O_2) indicates an increased entropy. At high temperatures, the effect of the entropy change becomes sufficient to overcome the effect of the energy change and the reaction proceeds. If the temperature is not high enough, the reaction will not take place.

Common Ion Effect

Adding more of one of the ions produced by a slightly soluble salt in solution results in precipitation of the salt. For example, in a system consisting of some silver chloride crystals in the bottom of a container of water, an equilibrium will be established between the excess of AgCl and its ions, Ag^+ and Cl^-.

$$AgCl_{(s)} \rightleftarrows Ag^+_{(aq)} + Cl^-_{(aq)}$$

Addition of NaCl to this system increases the $Cl^-_{(aq)}$ concentration. This causes a shift in equilibrium to the left, producing additional AgCl, which precipitates; The common ion effect, therefore, can be said to have *reduced* the solubility of the AgCl.

QUESTIONS

1. Which is the expression for the free energy change of a chemical reaction?   (1) $\Delta H = \Delta G - T\Delta S$ (2) $\Delta G = \Delta S - T\Delta H$ (3) $\Delta G = \Delta H - T\Delta S$ (4) $\Delta S = \Delta G - T\Delta H$

2. The change in free energy of a chemical reaction is represented by   (1) ΔT (2) ΔS (3) ΔH (4) ΔG

3. The free energy change, ΔG, must be negative when   (1) ΔH is positive and ΔS is positive   (2) ΔH is positive and ΔS is negative   (3) ΔH is negative and ΔS is positive   (4) ΔH is negative and ΔS is negative

4. In the free energy equation $\Delta G = \Delta H - T\Delta S$, the symbol T refers to (1) time in seconds (2) time in hours (3) Celsius temperature (4) Kelvin temperature

5. A sample of H_2O (ℓ) at 20° C is in equilibrium with its vapor in a sealed container. When the temperature increases to 25°C, the entropy of system will (1) decrease (2) increase (3) remain the same

6. Above 0°C, ice changes spontaneously to water according to the following equation: $H_2O(s) + heat \rightarrow H_2O(\ell)$
    The changes in $H_2O(s)$ involve
    (1) an absorption of heat and a decrease in entropy (2) a release of heat and a decrease in entropy (3) an absorption of heat and an increase in entropy (4) a release of heat and an increase in entropy

7. As 1 gram of H_2O (g) changes to 1 gram of H_2O (ℓ), the entropy of the system (1) decreases (2) increases (3) remains the same

8. Which reaction results in an increase in entropy?
    (1) $2H_2O(\ell) \rightarrow 2H_2(g) + O_2(g)$ (2) $2H_2(g) + O_2(g) \rightarrow 2H_2O(\ell)$
    (3) $N_2(g) + 3H_2(g) \rightarrow 2NH_3(g)$ (4) $2CO(g) + O_2(g) \rightarrow 2CO_2(g)$

9. A 1 gram sample of a substance has the greatest entropy when it is in the (1) solid state (2) liquid state (3) crystalline state (4) gaseous state

10. Given the system at equilibrium:

    $$AgCl(s) \rightleftarrows Ag^+(aq) + Cl^-(aq)$$

    When 0.1 M HCl is added to the system, the equilibrium will shift to the (1) right and the concentration of Ag^+ (aq) will decrease (2) right and the concentration of Ag^+ (aq) will increase (3) left and the concentration of Ag^+ (aq) will decrease (4) left and the concentration of Ag^+ (aq) will increase

11. Given the system at equilibrium:

    $$CaSO_4(s) \rightleftarrows Ca^{2+}(aq) + SO_4^{2-}(aq)$$

    If $K_2SO_4(s)$ is added and the temperature remains constant, the Ca^{2+}(aq) concentration will
    (1) decrease, and the amount of $CaSO_4$ (s) will decrease
    (2) decrease, and the amount of the $CaSO_4$ (s) will increase
    (3) increase, and the amount of $CaSO_4$ (s) will decrease
    (4) increase, and the amount of $CaSO_4$ (s) will increase

12. What is the free energy of formation of H_2O (ℓ) in kilocalories per mole at 1 atmosphere and 298 K?
    (1) −54.6   (2) − 56.7   (3) −57.8   (4) −68.3

13. The diagram below shows a system of gases with the valve closed.

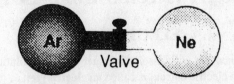

As the valve is opened, the entropy of the gaseous system (1) decreases (2) increases (3) remains the same

14. As the temperature of a system increases, the entropy of the system (1) decreases (2) increases (3) remains the same

15. Given the reaction at equilibrium:

$$NaCl(s) \quad \rightleftarrows \quad Na^+ (aq) + Cl^- (aq)$$

The addition of KCl to this system will cause a shift in the equilibrium to the
(1) left, and the concentration of $Na^+(aq)$ ions will increase
(2) right, and the concentration of $Na^+(aq)$ ions will increase
(3) left, and the concentration of $Na^+(aq)$ ions will decrease
(4) right, and the concentration of $Na^+(aq)$ ions will decrease

16. When a reaction has a negative ΔG, it must be (1) exothermic (2) endothermic (3) spontaneous (4) nonspontaneous

UNIT 7. ACIDS AND BASES

ELECTROLYTES

Any substance that will dissolve in water to form a solution capable of conducting an electric current is called an **electrolyte**.

In order to conduct electric current, a solution must contain ions that are free to move. *All* ionic compounds are electrolytes. *Some* polar covalent compounds will form ions and conduct electricity when dissolved in water. Examples include HCl and HBr. The explanation for the behavior of electrolytes was furnished by Svante Arrhenius.

Weak electrolytes in aqueous solutions reach equilibrium between the ions and the undissociated compound. The equilibrium constant for these systems is called the **dissociation constant**. (Note-dissociation means the separation of ions.) Dissociation constants change in value when the temperature changes, as do all equilibrium constants.

ACIDS AND BASES

In the processing of information, it is customary to define terms in two ways. One definition is based on experimental observations and includes a set of conditions to apply when a term is used in a particular situation. This is called an **operational definition**. The second definition is based on the interpretation of observed facts. These are called **conceptual definitions**. Conceptual definitions of acids and bases have been expanded to include additional information provided by greater understanding of how they behave, including reactions not involving aqueous solutions.

Acids
Operational Definitions of Acids

1. Aqueous solutions of acids are electrolytes. The ability of acids to conduct electricity is proportional to the extent to which they ionize. Some acids ionize almost completely in aqueous solution and are *strong electrolytes.* These are called the **strong acids**, and include HI, HBr, HCl, HNO_3, and H_2SO_4. Other acids are *weak electrolytes* because they ionize only to a slight degree. These are called the **weak acids**, and include CH_3COOH (acetic acid). Reference Table L lists the relative strengths of acids. The strong acids have large K_a values, the weak acids have small K_a values.

2. Acids react with certain metals to produce hydrogen gas and a salt. These metals are listed below hydrogen in the Reference Table N. (Standard Electrode Potentials). Certain acids, such as HNO_3, do not produce hydrogen gas when they react with metals except when the acids are very dilute. This is because the acids are strong oxidizing agents and, as such, react with metals in a different way.

3. Acids cause acid-base indicators to change in color. Acid-base indicators are substances whose color depends on the concentration of hydrogen ions. *Litmus,* a common indicator, is blue in basic solutions and red in acid solutions. *Phenolphthalein,* another indicator, is pink in basic solutions and colorless in acid solutions.

4. Acids react with hydroxides (compounds that produce OH^- ions) to form water and a salt. These reactions are known as **neutralization** reactions.

5. Acids, in dilute aqueous solutions, have a sour taste. This property is the basis for the sour taste of many common foods. The citrus fruits—oranges, lemons, etc.—contain citric acid. The characteristic taste of vinegar is due to the acetic acid it contains. Taste is *NOT* a property to be used as a test. **Never taste laboratory chemicals!**

Conceptual Definitions of Acids

1. The **Arrhenius Theory** of acids defines an acid as a substance that yields hydrogen ions (H^+) as the only positive ions in aqueous solution. Furthermore, the properties of acids in solution are caused by the excess of hydrogen ions.

2. The **Brönsted-Lowry Theory** of acids defines an acid as any species (molecule or ion) that can donate a proton to another species. Since hydrogen ions are protons, this theory does not replace Arrhenius' theory. It simply extends it to include all proton donors. For example, when ammonia (NH_3) is dissolved in water, the water donates a proton to the ammonia to produce the ammonium ion. The equation for this reaction is

$$NH_3 + H_2O \rightleftarrows NH_4^+ + OH^-$$

According to the Brönsted-Lowry theory, water is considered an acid in this reaction.

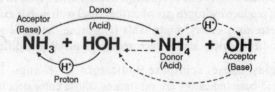

Figure 7-1. Water is an acid in this reaction.

Bases

Operational Definitions of Bases

1. Aqueous solutions of bases are electrolytes. Like acids, bases may be strong or weak, depending on their degree of ionization. This information is also provided in Reference Table L. Strong bases include sodium hydroxide (NaOH) and potassium hydroxide (KOH). Ammonium hydroxide (NH_4OH) is a weak base.

2. Bases cause acid-base indicators to change color. Adding base to a solution results in lowering the concentration of hydrogen ions.

3. Bases react with acids to form salts and water.

4. Aqueous solutions of bases feel slippery. Strong bases are caustic and can cause severe burns. Thus, this slippery feeling provides a warning. Whenever it is felt, the affected area should be thoroughly flushed with cold running water.

Conceptual Definitions of Bases

1. Arrhenius' theory defines a base as a substance that yields hydroxide ions ($-OH^-$) as the only negative ions in aqueous solution. According to Arrhenius' definition, the only bases are hydroxides and all properties of bases in aqueous solutions are caused by the hydroxide ion.

2. The Brönsted-Lowry theory defines a base as any species (molecule or ion) that can combine with or accept a proton. This definition extends Arrhenius' definition to include many species in addition to the hydroxides. For example, in the reaction of hydrochloric acid with water

$$HCl + H_2O \rightleftarrows H_3O^+ + Cl^-$$

the water molecule accepts a proton, forming a hydronium ion (H_3O+).
In this reaction, water is a base according to the Brönsted-Lowry theory.

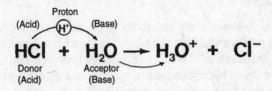

Figure 7-2. Water is a base in this reaction.

Conjugate Acid-Base Pairs

In an acid-base reaction (see above), the acid transfers a proton to the
base. The acid is then capable of accepting a proton. In effect, the particle
produced has become a base (a proton acceptor). The acid and its newly
formed base constitute a **conjugate acid-base pair.** Similarly, once a base
accepts a proton, it becomes capable of donating a proton as a conjugate
acid. In this way, a second conjugate acid-base pair is formed.
In the following, the two conjugate acid-base pairs are identified by
matching subscripts:

$$Acid_1 + Base_2 \rightleftarrows Base_1 + Acid_2$$

$$HCl + H_2O \rightleftarrows Cl^- + H_3O^+$$

$$H_2O + NH_3 \rightleftarrows OH^- + NH_4^+$$

Conjugate acid-base pairs are identified in Reference Table L (Relative
Strengths of Acids). The strongest acids have the weakest conjugate bases.
The strongest bases have the weakest conjugate acids.

Amphoteric (Amphiprotic) Substances

Certain substances can act either as acids or bases, depending on the
nature of the substance(s) with which they are interacting. According to
the Brönsted-Lowry theory, substances that can either donate or accept
protons, such as H_2O and HSO_4^-, are referred to as amphoteric or
amphiprotic. Any substance listed on both sides of Reference Table L is
amphoteric.

ACID-BASE REACTIONS

Neutralization

Reactions between acids and bases to produce salts and water are referred to as **neutralization** reactions. The term neutralization actually applies to the reaction in which one mole of hydrogen ions combines with one mole of hydroxide ions to produce one mole of water. Such reactions occur when equivalent quantities of acids and hydroxides are mixed.

Look at the equation for the reaction between hydrochloric acid, HCl, and sodium hydroxide, NaOH:

$$HCl + NaOH \rightarrow NaCl + H_2O$$

The neutralization portion of this reaction occurs between hydrogen ions from the acid and hydroxide ions from the base:

$$H^+ + OH^- \rightarrow HOH \text{ (water)}$$

The Cl^- ions from the acid and the Na^+ ions from the base are not part of the neutralization. They are merely **"spectator" ions**.

Acid-base titration. The molarity of an acid (or base) of unknown concentration can be determined in the lab by slowly adding it to a measured volume of a base (or acid) until neutralization occurs. This process is known as **titration**. The **end point** of a titration occurs when neutralization has just been attained. This end point can be determined by changes in the color of acid-base indicators, changes in temperature, or changes in electrode potential.

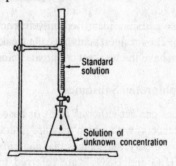

Standard solution

Solution of unknown concentration

Figure 7.3 Titration.

Calculations of the molarity of unknown concentration depend on the relationship between moles and volume and on the fact that the moles of H^+ must equal the moles of OH^-.

QUESTIONS

1. Which substance is an electrolyte? (1) C_2H_5OH (2) $C_6H_{12}O_6$ (3) $C_{12}H_{22}O_{11}$ (4) CH_3COOH

2. Which of the following is the best conductor of electricity? (1) NaCl(s) (2) NaCl (aq) (3) $C_6H_{12}O_6$(s) (4) $C_6H_{12}O_6$(aq)

3. Which of the following 0.1 M solutions is the best conductor of electricity? (1) H_2S(aq) (2) HCl(aq) (3) $C_6H_{12}O_6$(aq) (4) $C_{12}H_{22}O_{11}$(aq)

4. Which type of reaction will occur when equal volumes of 0.1 M HCl and 0.1 M NaOH are mixed? (1) neutralization (2) ionization (3) electrolysis (4) hydrolysis

5. The OH^- ion concentration is greater than the H_3O^+ ion concentration in a water solution of (1) CH_3OH (2) $Ba(OH)_2$ (3) HCl (4) H_2SO_4

6. Which is a characteristic of an aqueous solution of HNO_3? (1) It conducts electricity. (2) It forms OH^- ions. (3) It turns litmus blue. (4) It turns phenolphthalein pink.

7. Which solution will change litmus from blue to red? (1) NaOH(aq) (2) NH_4OH(aq) (3) CH_3OH(aq) (4) CH_3COOH(aq)

8. Which solution will turn litmus from red to blue? (1) H_2S(aq) (2) NH_3(aq) (3) SO_2(aq) (4) CO_2(aq)

9. Which metal will react with hydrochloric acid to yield hydrogen gas as one of the products? (1) Cu (2) Ca (3) Ag (4) Hg

10. Which substance is always produced in the reaction between hydrochloric acid and sodium hydroxide? (1) water (2) hydrogen gas (3) oxygen gas (4) a precipitate

11. Which compound reacts with an acid to form a salt and water? (1) CH_3Cl (2) CH_3COOH (3) KCl (4) KOH

12. Which equation represents a neutralization reaction?
    (1) H^+(aq) + OH^-(aq) $\rightarrow$ H_2O (ℓ)
    (2) Ag^+(aq) + I^-(aq) $\rightarrow$ AgI (s)
    (3) Zn (s) + 2HCl (aq) $\rightarrow$ $ZnCl_2$ (aq) + H_2 (g)
    (4) NaCl (aq) + $AgNO_3$ (aq) $\rightarrow$ $NaNO_3$ (aq) + AgCl (s)

13. Based on Reference Table E, which saturated solution is most basic? (1) potassium hydroxide (aq) (2) zinc hydroxide (aq) (3) calcium hydroxide (aq) (4) aluminum hydroxide (aq)

14. According to the Arrhenius theory, when a base is dissolved in water it will produce a solution containing only one kind of negative ion. To which ion does the theory refer? (1) hydride (2) hydroxide (3) hydrogen (4) hydronium

15. When an Arrhenius acid is dissolved in water, it produces (1) H^+ as the only positive ion in solution (2) NH_4^+ as the only positive ion in solution (3) OH^- as the only negative ion in solution (4) HCO_3^- as the only negative ion in solution

16. Which species is classified as an Arrhenius base (1) CH_3OH (2) $LiOH$ (3) PO_4^{3-} (4) CO_3^{2-}

17. According to Reference Table L, which of the following Brönsted acids has the strongest conjugate base? (1) HF (2) HCl (3) HBr (4) HI

18. According to Reference Table L, which ion can act as a Brönsted acid? (1) a sulfate ion (2) a hydrogen sulfate ion (3) a sulfide ion (4) a sulfite ion

19. Given the reaction:
$$HX + H_2O \rightarrow H_3O^+ (aq) + X^- (aq)$$
Based on the equation, HX would be classified as (1) a base, because it donates a proton (2) a base, because it accepts a proton (3) an acid, because it donates a proton (4) an acid, because it accepts a proton

20. In the reaction $NO_2^- (aq) + H_2O(\ell) \rightarrow HNO_2(aq) + OH^-(aq)$, the $NO_2^- (aq)$ acts as (1) a Brönsted acid (2) a Brönsted base (3) an Arrhenius acid (4) an Arrhenius base

21. Which equation illustrates H_2O acting as a Brönsted-Lowery base?

(1) $H^+(aq) + H_2O \rightarrow H_3O^+ (aq)$

(2) $CH_3COO^-(aq) + H_2O \rightarrow CH_3COOH (aq) + OH^- (aq)$

(3) $2Na + 2H_2O \rightarrow 2NaOH (aq) + H_2$

(4) $C + H_2O \rightarrow CO + H_2$

22. In the reaction $HSO_4^- + H_2O \rightleftarrows H_3O^+ + SO_4^{2-}$, an acid-base conjugate pair is (1) HSO_4^- and SO_4^{2-} (2) HSO_4^- and H_2O (3) SO_4^{2-} and H_3O^+ (4) SO_4^{2-} and H_2O

23. Given the equation: $NH_3(g) + H_2O(\ell) \rightleftarrows NH_4^+(aq) + OH^- (aq)$ The two Brönsted-Lowry acids are (1) NH_3 and H_2O (2) NH_3 and NH_4^+ (3) H_2O and NH_4^+ (4) H_2O and OH^-

24. In the reaction $HS^- + NH_2^- \rightleftarrows NH_3 + S^{2-}$, the two Brönsted bases are (1) HS^- and NH_2^- (2) HS^- and NH_3 (3) S^{2-} and NH_2^- (4) S^{2-} and NH_3

25. According to Reference Table L, which species is amphiprotic? (1) NH_4^+ (2) NH_2^- (3) HS^- (4) S^{2-}

26. Given the reactions:
    (A) $NH_3(g) + H_2O(\ell) \rightarrow NH_4^+(aq) + OH^-(aq)$
    (B) $HCl(aq) + H_2O(\ell) \rightarrow H_3O^+(aq) + Cl^-(aq)$
    As shown in equations (A) and (B) and based on the Brönsted theory, water is an amphoteric substance because it can (1) donate protons, only (2) accept protons, only (3) either donate or accept protons (4) neither donate nor accept protons

27. How many milliliters of 0.2 M NaOH are required to exactly neutralize 40 milliliters of 0.1 M HCl? (1) 10 (2) 20 (3) 40 (4) 80

28. How many milliliters of a 4.0-molar solution of HCl are needed to completely neutralize 60. milliliters of a 3.2-molar solution of NaOH? (1) 24 mL (2) 48 mL (3) 60. mL (4) 75 mL

29. Given the reaction: $HF + H_2O \rightleftarrows F^- + H_3O^+$
    Which species is the Brönsted acid in the reverse reaction (1) HF (2) H_2O (3) F^- (4) H_3O^+

30. According to Reference Table L, which is the strongest Brönsted base? (1) I^- (2) F^- (3) NH_2^- (4) OH^-

31. In the reaction $H_2O + H_2O \rightarrow H_3O^+ + OH^-$, the water is acting as (1) a proton acceptor, only (2) a proton donor, only (3) both a proton acceptor and a proton donor (4) neither a proton acceptor nor a proton donor

32. If 50. milliters of 0.50 M HCl is used to completely neutralize 25 milliliters of KOH solution, what is the molarity of the base? (1) 1.0 M (2) 0.25 M (3) 0.50 M (4) 2.5 M

IONIZATION CONSTANT

The equilibrium constant for an ionization reaction is called the **ionization constant.** The ionization constant for acids (K_a) is used to compare the relative strengths of acids. The higher the ionization constant, the stronger the acid.

Acids that ionize completely are very strong. However, because the concentration of undissociated acid approaches zero, K_a approaches infinity and cannot be calculated. In such cases, the K_a is listed as "very large" in the reference tables.

Ionization constants can be calculated for all acids that do not dissociate completely. Consider the dissociation reaction for acid HY:

$$HY \rightleftarrows H^+ + Y^-$$

The ionization constant for this reaction is

$$K_a = \frac{[H^+][Y^-]}{[HY]}$$

If the acid dissociated completely, the concentration of XY would be zero. In the expression for K_a, the denominator would be zero and the K_a value would be infinite. K_a's are listed on Reference Table L. **For weak acids the higher the K_a, the stronger the acid.**

K_w, The Ionization Constant for Water

The equation for the ionization of water is

$$H_2O(\ell) \rightarrow H^+(aq) + OH^-(aq)$$

The ionization constant for this reaction is

$$K_a = [H^+][OH^-]$$

In pure water at 25°C, $[H^+] = 1 \times 10^{-7}$ moles per liter. As the equation indicates, an equal number of moles of OH^- are produced. Thus, $[OH^-]$ is also 1×10^{-7} moles per liter. Since the concentration of water molecules remains virtually unchanged in all reactions in aqueous solutions, water is not included in an ionization constant expression. The ionization constant of water is given the expression K_w:

$$K_w = [H^+][OH^{-1}] = 1 \times 10^{-14}$$

This value of K_w provides a useful tool in calculations involving $[H^+]$ and $[OH^-]$. If the concentration of either ion is known, the concentration of the other can be calculated as:

Concentration of unknown ion $= (1 \times 10^{-14}) \div$ concentration of the known ion

For example, in an aqueous solution containing 1×10^{-12} moles per liter of hydroxide ions, the hydrogen ion concentration, $[H^+]$, will be $(1 \times 10^{-14}$ moles per liter) $\div (1 \times 10^{-12}$ moles per liter). Thus, $[H^+] = 1 \times 10^{-2}$ moles per liter. Since the dominant ion in this solution is the hydrogen ion $(1 \times 10^{-2}$ moles per liter), the solution is acidic.

pH

Expressing hydrogen ion concentration in terms of moles per liter is rather cumbersome. A more convenient method, called **pH values,** has been devised. The pH is the logarithm, or log, of the reciprocal of the hydrogen ion concentration.

The log of a value is the power to which 10 must be raised to equal that value. The log of the reciprocal is the negative log of that value. Whole number values of pH can be found by simply changing the exponent of the hydrogen ion concentration to a positive number.

For example, the pH value equivalent to a hydrogen ion concentration of 1×10^{-7} moles per liter is 7. Pure water is considered to have a pH of 7 at 25°C. A pH less than 7 indicates a higher value for hydrogen ion concentration. Solutions having a pH less than 7 are acidic. Those with pH values greater than 7 are basic.

To illustrate this idea using logarithms:

$$pH = \log \frac{1}{[H^+]} = -\log [H^+]$$

$$\text{When } [H^+] = 1 \times 10^{-7}$$
$$pH = -\log (1 \times 10^{-7})$$
$$= -(-7) = 7$$

It is possible to find the hydrogen ion concentration if the pH of a solution is known:

$$[H^+] = 1 \times 10^{-pH}$$

Given that the pH of a solution is 6, then $[H^+]$ of that solution is 1×10^{-6}. It is also possible to determine the hydroxide ion concentration of a solution of known pH. For example, to determine $[OH^-]$ of a solution with pH of 5 at 25°C, first find $[H^+]$ of the solution:

$$[H^+] = 1 \times 10^{-pH} = 1 \times 10^{-5}$$

Next, use the expression $[H^+] [OH^-] = 1 \times 10^{-14}$ to find $[OH^-]$:

$$[OH^-] = \frac{1 \times 10^{-14}}{[H^+]}$$
$$= \frac{1 \times 10^{-14}}{1 \times 10^{-5}}$$
$$= 1 \times 10^{-9}$$

SAMPLE PROBLEM

What is the pH of a solution of 0.001 M hydrochloric acid?
Solution:

Since HCl is a strong acid, ionization is virtually 100%, and $[H^+]$ is the same as the molarity of the solution: 1×10^{-3} M.

$$pH = 3.0$$

---

pOH

The definition of pOH is similar to pH and applies to the OH^- ion.

$$pOH = \log \frac{1}{\left[OH^-\right]} = -\log\left[OH^-\right]$$

pH and pOH are related as follows:

$$\mathbf{pH + pOH = 14}$$

What is the pOH of the HCl solution described in the preceding problem?

Solution:
$$pH + pOH = 14.0$$
$$pOH = 14.0 - 3.0 = \mathbf{11.0}$$

What is the hydrogen ion concentration of a solution that has a pOH of 5.0?

Solution:
$$pH = 14.0 - pOH$$
$$= 14.0 - 5.0 = 9.0$$
$$-\log [H^+] = pH = 9.0$$
$$\log [H^+] = -9.0$$
$$[H^+] = \text{antilog } (-9.0) = \mathbf{1 \times 10^{-9}}$$

---

pH and Salts (Hydrolysis)

Salts. Ionic compounds containing positive ions other than the hydrogen ion and negative ions other than the hydroxide ion are called salts. Most salts dissociate completely into their ions when dissolved in water. Salts are, therefore, strong electrolytes. Some salts react with water to form solutions that have properties of acids (acidic solutions), while others form solutions that have properties of bases (basic solutions). The process responsible for this behavior is called **hydrolysis**.

All salts are products of neutralization reactions between acids and bases. Salts derived from the neutralization of weak acids by strong bases will produce basic solutions. For example, sodium carbonate, Na_2CO_3, is formed by the neutralization of carbonic acid (a weak acid) by sodium hydroxide (a strong base). When the salt is dissolved in water, the solution is basic:

$$Na_2CO_3 + 2H_2O \rightleftarrows H_2CO_3 + 2NaOH$$

The dissociation of the NaOH produces 100% hydroxide ions, OH^-, which is characteristic of a strong base.

Salts formed by the neutralization of a weak base by a strong acid will produce acidic solutions. Ammonium chloride, NH_4Cl, a salt, is produced by the neutralization of ammonium hydroxide (a weak base) by hydrochloric acid (a strong acid). When dissolved in water, ammonium chloride produces an acidic solution:

$$NH_4Cl + H_2O \rightleftarrows HCl + NH_4OH$$

The dissociation of the Ha produces 100% H^+ ions, which is a characteristic of a strong acid.

Apparently hydrolysis does not occur when salts formed from strong acids and strong bases dissolve in water, therefore the solution will be neutral.

QUESTIONS

1. What is the pH of a solution whose H^+ ion concentration is 0.0001 mole per liter?   (1) 1   (2) 10   (3) 14   (4) 4
2. In a solution with a pH of 3, the color of (1) litmus is red (2) litmus is blue (3) phenolphthalein is red (4) phenolphthalein is blue
3. What is the concentration of H^+ ions, in moles per liter, of a 0.0001 M HCl solution?
   (1) 1×10^{-1} (2) 1×10^{-2} (3) 1×10^{-3} (4) 1×10^{-4}
4. The ionization constants (K_a's) of four acids are shown below. Which K_a represents the *weakest* of these acids?

   (1) $K_a = 1.0 \times 10^{-5}$ (2) $K_a = 1.0 \times 10^{-4}$ (3) $K_a = 7.1 \times 10^{-3}$
   (4) $K_a = 1.7 \times 10^{-2}$

5. What is the ionization constant, K_a, for H_2S at 1 atmosphere and 298 K?
   (1) 1.0×10^{-4} (2) 1.0×10^{-5} (3) 9.5×10^{-8} (4) 1.0×10^{-14}

6. The reaction $CH_3COOH(aq) \rightleftarrows CH_3COO^-(aq) + H^+(aq)$ has a K_a equal to 1.8×10^{-5} at 25° C. In a solution of this acid at 25° C, the concentration of CH_3COOH is
   (1) less than the concentration of H^+ ions
   (2) equal to the concentration of H^+ ions
   (3) greater than the concentration of CH_3COO^- ions
   (4) equal to the concentration of CH_3COO^- ions

7. The Kw value for a sample of water at 1 atmosphere and 298 K will be most likely to change when there is an increase in the
   (1) concentration of H^+ ions (2) concentration of OH^- ions
   (3) pressure (4) temperature

8. What is the pH of a solution that has a hydrogen ion concentration of 1×10^{-10} mole per liter? (1) 1 (2) 10 (3) 14 (4) 4

9. What is the ionization constant (K_a) of acetic acid at 298 K?
   (1) 6.0×10^{-23} (2) 1.0×10^{-14} (3) 2.2×10^{-7} (4) 1.8×10^{-5}

10. What is the hydroxide ion concentration of a solution that has a hydronium ion concentration of 1×10^{-9} mole per liter at 298 K?
    (1) 1×10^{-5} mole per liter   (2) 1×10^{-7} mole per liter
    (3) 1×10^{-9} mole per liter   (4) 1×10^{-14} mole per liter

11. What is the $[H^+]$ of a 0.001 M NaOH solution?
    (1) 1×10^{-1} (2) 1×10^{-7} (3) 1×10^{-11} (4) 1×10^{-14}

12. Which concentration indicates a basic solution at 298K?
    (1) $[OH^-] > 1.0 \times 10^{-7}$ (2) $[OH^-] = 1.0 \times 10^{-7}$
    (3) $[H_3O^+] > 1.0 \times 10^{-7}$ (4) $[H_3O^+] = 1.0 \times 10^{-7}$

13. What is the H^+ ion concentration of an aqueous solution that has a pH of 11?
    (1) 1.0×10^{-11} mol/L   (2) 1.0×10^{-3} mol/L
    (3) (1) 3.0×10^{-1} mol/L   (4) 11×10^{-1} mol/L

14. If a solution has a hydronium ion concentration of 1×10^{-9} M, the solution is (1) basic and has a pH of 9 (2) basic and has a pH of 5 (3) acidic and has a pH of 9 (4) acidic and has a pH of 5

15. Which salt will hydrolize in water to produce a basic solution? (1) BaI_2 (2) $NaNO_2$ (3) $CaCl_2$ (4) $MgSO_4$

16. The $[OH^-]$ of a solution is 1×10^{-6}. At 1 atmosphere and 298 K, the product of the $[H^+][OH^-]$ is (1) 1×10^{-2} (2) 1×10^{-6} (3) 1×10^{-8} (4) 1×10^{-14}

17. As additional solid KCl is added to a saturated solution of KCl, the conductivity of the solution (1) decreases (2) increases (3) remains the same

18. When equal volumes of 0.5 M HCl and 0.5 M NaOH are mixed, the pH of the resulting solution is (1) 1 (2) 2 (3) 7 (4) 4

19. Adding 0.1 M NaOH to a 0.1 M solution of HCl will cause the pH of the solution to (1) decrease (2) increase (3) remain the same

20. The $[H^+]$ of a solution is 1×10^{-2} at 298 K. What is the $[OH^-]$ of this solution? (1) 1×10^{-14} (2) 1×10^{-12} (3) 1×10^{-7} (4) 1×10^{-2}

21. As a solution of NaOH is diluted from 0.1 M to 0.001 M, the pH of the solution (1) decreases (2) increases (3) remains the same

22. Which 0.1M solution has the highest concentration of H_3O^+ ions? (1) CH_3COOH (2) NaCl (3) KBr (4) $Ba(OH)_2$

23. An indicator was used to test a water solution with a pH of 12. Which indicator color would be observed? (1) colorless with litmus (2) red with litmus (3) colorless with phenolphthalein (4) pink with phenolphthalein

24. What is the hydroxide ion concentration of a solution with a pH of 4?
    (1) 1×10^{-14} mole/liter (2) 1×10^{-10} mole/liter
    (3) 1×10^{-7} mole/liter (4) 1×10^{-4} mole/liter

25. What is the pH of the solution formed by completely neutralizing 50 milliliters of 0.1 M HNO_3 with 50 milliliters of 0.1 M NaOH at 298 K? (1) 1 (2) 7 (3) 10 (4) 4

26. What color is phenolphthalein in a solution that has a pH of 9? (1) blue (2) pink (3) white (4) colorless

27. A 0.1 M acid solution at 298 K would conduct electricity best if the acid had a K_a value of (1) 1.0×10^{-7} (2) 1.8×10^{-5} (3) 6.7×10^{-4} (4) 1.7×10^{-2}

28. The reaction of NH_4NO_3 with water to form an acidic solution is called (1) oxidation (2) reduction (3) electrolysis (4) hydrolysis

29. Which type of reaction is represented by the following equation?

$$Al_2S_3 + 6H_2O \rightarrow 2Al(OH)_3 + 3H_2S$$

(1) neutralization (2) dehydration (3) electrolysis (4) hydrolysis

30. As the concentration of NH_4Cl in a solution increases, the pH of the solution (1) decreases (2) increases (3) remains the same

31. Which could be the pH of a solution whose H^+ ion concentration is less than the OH^- ion concentration? (1) 9 (2) 2 (3) 3 (4) 4

32. Which solution would turn red litmus paper blue? (1) NaCl (aq) (2) HCl (aq) (3) $NaC_2H_3O_2$ (aq) (4) $HC_2H_3O_2$ (aq)

33. An aqueous solution of $NaC_2H_3O_2$ is basic. The salt $NaC_2H_3O_2$ can be derived from the reaction of a (1) strong acid with a weak base (2) strong acid with a strong base (3) weak acid with a weak base (4) weak acid with a strong base

34. Which salt hydrolyzes in water to form a solution that is acidic? (1) KCI (2) NaCI (3) NH_4Cl (4) LiCI

35. When K_2CO_3 is dissolved in water, the resulting solution turns litmus paper (1) red and is acidic (2) blue and is acidic (3) red and is basic (4) blue and is basic

36. A 0.1 M solution of which acid is the best conductor of electricity at 25°C?

(1) H_3PO_4 ($K_a = 7.5 \times 10^{-3}$)   (2) HNO_2 ($K_a = 4.6 \times 10^{-4}$)
(3) CH_3COOH ($K_a = 1.8 \times 10^{-5}$)   (4) H_2S ($K_a = 1.0 \times 10^{-7}$)

UNIT 8.
REDOX AND ELECTROCHEMISTRY

REDOX

Redox is a term used for oxidation-reduction reactions. The single term redox is useful because both processes—oxidation and reduction—always occur together in the same reaction. Redox reactions result from the competition for electrons between atoms.

Oxidation

Oxidation is defined as the loss, or apparent loss, of electrons. The term is used in referring to any chemical change in which there is an increase in oxidation number. The particle that experiences the loss of electrons (and increase in oxidation number) is said to have been *oxidized.* Any such particle is called a **reducing agent.**

Reduction

Reduction is defined as the gain, or apparent gain, of electrons, with an accompanying decrease in oxidation number. The particle that experiences the gain of electrons (and decreased oxidation number) is said to have been *reduced.* Since this particle oxidizes another particle by removing an electron from it, this particle is called an **oxidizing agent.**

The terms oxidation, reduction, reducing agent, and oxidizing agent can be clarified by describing the events in the following reaction:

$$Zn(s) + 2H^+(aq) \rightarrow Zn^{2+}(aq) + H_2(g)$$

Zn loses electrons and is oxidized to Zn^{2+}. Zn is the reducing agent for H^+.

H^+ gains electrons and is reduced to H_2. H^+ is the oxidizing agent for Zn.

Oxidation Number (Oxidation State)

Oxidation numbers are assigned to atoms or ions as convenient devices to keep track of electron transfers. The assignment is based on the arbitrary assumption that electrons shared between two atoms belong to the atom with the higher electronegativity. The oxidation number can be defined as a fictitious charge assigned to an atom or ion on the basis of a certain set of rules.

Rules for Assigning Oxidation Numbers

1. The oxidation number of each atom in free elements is always zero. The hydrogen in H_2, the sodium in Na, and the sulfur in S_8 all have oxidation numbers of zero.

2. The oxidation numbers of ions is the same as the charge on the ion. In $MgCl_2$ the Mg^{2+} ion has an oxidation number of +2. Each of the two Cl^- ions has an oxidation number of −1. In $FeCl_2$ the Fe^{2+} ion has an oxidation number of +2, while in $FeCl_3$ the Fe^{3+} ion has an oxidation number of +3.

Special Note: It should be noted that when writing symbols for the charge on an ion, the numeral is written *before* the + or − sign (2+, 3+, 2−, etc.). When writing oxidation numbers, the + or − sign precedes the numeral (+1, +3, −1, etc.).

3. Since the Group 1 metals form only 1 + ions, their oxidation number is + 1 in all their compounds.

4. The metals in Group 2 form only 2+ ions, and their oxidation number is +2 in all their compounds.

5. Oxygen has a −2 oxidation number in practically all of its compounds. This rule is especially useful when identifying the oxidation numbers of elements in polyatomic ions. The few exceptions for oxygen are in peroxides, such as H_2O_2 and Na_2O_2, where the oxidation number is − 1 and in compounds with fluorine (OF_2), in which it is +2.

6. Hydrogen has an oxidation number of + 1 in all its compounds *except* metal hydrides formed with Group 1 and Group 2 metals (such as LiH and CaH_2 in which it is − 1).

7. The sum of the oxidation numbers in a compound must be zero.

8. The sum of the oxidation numbers in a polyatomic ion must be equal to the charge on the ion. In the CO_3^{2-} ion the three oxygens provide a total of −6. Since there is one carbon, it must contribute +4 in order for CO_3^{2-} to have a net oxidation number of −2.

Finding Oxidation Numbers

The following will illustrate practical methods of finding oxidation numbers.

1. Find the oxidation numbers of N and O in the compound N_2O. Using rule 2, oxygen has an oxidation number of − 2. Using rule 7, the total oxidation number of the nitrogen is +2. Thus, each nitrogen atom has an oxidation number of +1.

2. For compounds composed of more than two elements, such as $KMnO_4$:

   a. Construct a table similar to the one shown here.

   b. Use rule 3 for K and rule 5 for O. Write the oxidation numbers for these elements in the table.

	K	Mn	O_4
	+1	?	−2
	+1	+7	−8

   c. Multiply the oxidation numbers by the subscripts. Write the products in the bottom row of the table.

   d. Apply rule 7. The total oxidation number of Mn is +7. Since there is only 1 atom of Mn, the oxidation number for Mn in this compound is +7.

QUESTIONS

1. Oxidation-reduction reactions occur because of the competition between particles for (1) neutrons (2) electrons (3) protons (4) positrons

2. All redox reactions involve (1) the gain of electrons, only (2) the loss of electrons, only (3) both the gain and the loss of electrons (4) neither the gain nor the loss of electrons

3. Which statement describes what occurs in the following redox reaction?

$$Cu(s) + 2Ag^+ (aq) \rightarrow Cu^{2+} (aq) + 2Ag (s)$$

(1) Only mass is conserved. (2) Only charge is conserved (3) Both mass and charge are conserved (4) Neither mass nor charge is conserved

4. In which compound does chlorine have the highest oxidation number? (1) KClO (2) $KClO_2$ (3) $KClO_3$ (4) $KClO_4$

5. In which compound does sulfur have an oxidation number of −2? (1) SO_2 (2) SO_3 (3) Na_2S (4) Na_2SO_4

6. In which compound does chlorine have an oxidation number of +7? (1) $HClO_4$ (2) $HClO_3$ (3) $HClO_2$ (4) HClO

7. In the reaction $2CrO_4^{2-} (aq) + 2H^+ (aq) \rightarrow Cr_2O_7^{2-} (aq) + H_2O(\ell)$, the oxidation number of chromium (1) decreases (2) increases (3) remains the same

8. Which is a redox reaction?
   (1) $Mg + 2HCl \rightarrow MgCl_2 + H_2$
   (2) $Mg(OH)_2 + 2HCl \rightarrow MgCl_2 + 2H_2O$
   (3) Mg^{2+} (aq) $+ 2OH^-$ (aq) $\rightarrow Mg(OH)_2$
   (4) $MgCl_2 + 6H_2O \rightarrow MgCl_2 \cdot 6H_2O$

9. In the reaction $Cl_2 + H_2O \rightarrow HClO + HCl$, the hydrogen is
   (1) oxidized, only (2) reduced, only (3) both oxidized and
   reduced (4) neither oxidized nor reduced

10. In the reaction $Zn + Fe^{2+} \rightarrow Zn^{2+} + Fe$, the reducing agent is
    (1) Zn (2) Fe^{2+} (3) Zn^{2+} (4) Fe

11. Which change in oxidation number represents reduction?
    (1) –1 to +1   (2) –1 to –2   (3) –1 to +2   (4) –1 to 0

12. As an S^{2-} ion is oxidized to an S^0 atom, the number of protons in its
    nucleus (1) decreases (2) increases (3) remains the same

13. In the reaction $4NH_3 + 5O_2 \rightarrow 4NO + 6H_2O$, the oxidation number
    of nitrogen changes from (1) –2 to–3 (2) –2 to +3 (3) –3 to–2
    (4) –3 to +2

14. Which is an oxidation-reduction reaction?
    (1) $4Na + O_2 \rightarrow 2Na_2O$ (2) $3O_2 \rightarrow 2O_3$
    (3) $AgNO_3 + NaCl \rightarrow AgCl + NaNO_3$ (4) $KI \rightarrow K^+ + I^-$

15. If element X forms the oxides XO and X_2O_3, the oxidation numbers
    of element X are (1) +1 and +2 (2) +2 and +3 (3) +1 and +3
    (4) +2 and +4

16. Given the oxidation-reduction reaction:

    $$H_2 + 2Fe^{3+} \rightarrow 2H^+ + 2Fe^{2+}$$

    Which species undergoes reduction? (1) H_2 (2) Fe^{3+} (3) H^+
    (4) Fe^{2+}

17. In the reaction $4Zn + 10HNO_3 \rightarrow 4Zn(NO_3)_2 + NH_4NO_3 + 3H_2O$,
    the zinc is (1) reduced and the oxidation number changes from 0
    to +2 (2) oxidized and the oxidation number changes from 0 to +2
    (3) reduced and the oxidation number changes from +2 to 0
    (4) oxidized and the oxidation number changes from +2 to 0

18. Which is a redox reaction?
    (1) $2KBr + F_2 \rightarrow 2KF + Br_2$
    (2) $2HCl + Mg(OH)_2 \rightarrow 2HOH + MgCl_2$
    (3) $2NaCl + H_2SO_4 \rightarrow Na_2SO_4 + 2HCl$
    (4) $Ca(OH)_2 + Pb(NO_3)_2 \rightarrow Ca(NO_3)_2 + Pb(OH)_2$

19. In the reaction $Zn(s) + Cu^{2+}$ (aq) $\rightarrow Zn^{2+}$ (aq) $+ Cu(s)$, the reducing
    agent is (1) Zn(s) (2) Cu(s) (3) Cu^{2+}(aq) (4) Zn^{2+}(aq)

20. In the reaction $2Al + 3Ni(NO_3)_2 \rightarrow 2Al(NO_3)_3 + 3Ni$, the aluminum is (1) reduced and its oxidation number increases (2) reduced and its oxidation number decreases (3) oxidized and its oxidation number increases (4) oxidized and its oxidation number decreases

21. Given the redox reaction: $Ni + Sn^{4+} \rightarrow Ni^{2+} + Sn^{2+}$ Which species has been oxidized? (1) Ni (2) Sn^{4+} (3) Ni^{2+} (4) Sn^{2+}

22. In which substance is the oxidation number of nitrogen zero? (1) N_2 (2) NH_3 (3) NO_2 (4) N_2O

23. What is the oxidation number of Pt in K_2PtCl_6? (1) -2 (2) $+2$ (3) -4 (4) $+4$

24. In the reaction $2H_2S + 3O_2 \rightarrow 2SO_2 + 2H_2O$, the oxidizing agent is (1) oxygen (2) water (3) sulfur dioxide (4) hydrogen sulfide

25. In the reaction $2Na + 2H_2O \rightarrow 2Na^+ + 2OH^- + H_2$, the substance oxidized is (1) H_2 (2) H^+ (3) Na (4) Na^+

26. Given the reaction: $3Cu + 8HNO_3 \rightarrow 3Cu(NO_3)_2 + 2NO + 4H_2O$ The reducing agent is (1) Cu^0 (2) N^{+5} (3) Cu^{+2} (4) N^{+2}

27. Given the reaction: $Sn^{2+}(aq) + 2Fe^{3+}(aq) \rightarrow Sn^{4+}(aq) + 2Fe^{2+}(aq)$ The oxidizing agent in this reaction is (1) Sn^{2+} (2) Fe^{3+} (3) Sn^{4+} (4) Fe^{2+}

ELECTROCHEMISTRY

Half-reactions

All redox reactions consist of two "parts"—a gain of electrons (reduction) and a loss of electrons (oxidation). Each of these "parts" of a redox reaction can be thought of as a half-reaction. A separate equation, including the gain or loss of electrons, can be written for each half-reaction.

For example, consider the reaction

$$Mg + Cl_2 \rightarrow MgCl_2$$

Written as half reactions, this reaction would be

oxidation: $Mg^\circ \rightarrow Mg^{2+} + 2e^-$
reduction: $Cl_2 + 2e^- \rightarrow 2Cl^-$

Half-cells

A **half-cell** is produced when a metal is placed in a solution of a salt of the metal. This arrangement provides a source of metal ions to be reduced and a source of metal atoms to be oxidized.

Electrochemical Cells

Half-cells of two different metals can be used to produce electricity, provided the metals are connected by a wire to carry the electrons and the solutions are brought into contact by a salt bridge to carry the ions without mixing the solutions. Such a system is called an electrochemical cell.

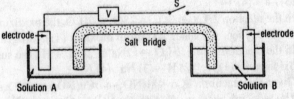

Figure 8-1.

In electrochemical cells, a spontaneous chemical reaction produces electricity

Electrolytic Cells

In an electrolytic cell, electricity is used to produce a non spontaneous chemical reaction. The reaction is called electrolysis. The energy required to proceed is usually supplied by an externally applied electric current.

The diagram below shows the electrolysis of fused (melted) sodium chloride.

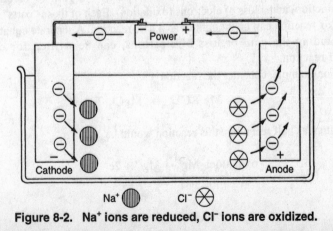

Figure 8-2. Na^+ ions are reduced, Cl^- ions are oxidized.

The overall molecular reaction occurring in this cell can be represented as follows:

$$2NaCl \rightarrow 2Na + Cl_2$$

The electrolysis of fused sodium chloride is the only process for the production of sodium metal.

Cell Potential (Voltage)

In electrochemical cells, the voltage produced represents the difference in potential of the two half-cells. There is no practical way of measuring the voltage associated with a single half-cell. Thus, in order to provide data for a scale of voltages, it was necessary to select a *standard half-cell* to which all other half-cells could be compared. The half-cell selected was the $H°/H^{1+}$ half-cell. This half-cell is assigned a potential of zero volts. The voltage produced by any other half-cell will be recorded as the voltage produced (measured by a voltmeter) when that half-cell is combined with a hydrogen half-cell to produce a current under standard conditions of concentration, temperature and pressure.

STANDARD ELECTRODE POTENTIALS (E°)

The voltages obtained by comparing the driving force of a half-reaction with that of a hydrogen half-cell provide the data for the table of Standard Electrode Potentials (Reference Table N).

Standard electrode potentials are useful in predicting whether or not a specific redox reaction will take place. A good rule to use in reading Reference Table N is: Any species on the left of the yield sign (reduction) will react with any species below it on the right of the yield sign (oxidation).

$$Cr^{3+} + Mn(s) \rightarrow \text{reaction}$$
$$Cr^{3+} + Ni(s) \rightarrow \text{no reaction}$$

Any pair of half-reactions in the table can be combined to give the complete reaction in an electrochemical cell. It must be noted that all the half-reactions in the tables are written as reductions. When combining half-reactions, one of the two must be an oxidation. The oxidation half-reaction must be written as the reverse of the one written in the table, and the sign of E° changed.

In the reaction $Mg + Cl_2 \rightarrow MgCl_2$, the chlorine is reduced. Thus, the chlorine half-reaction is written just as it appears in the table:

$$Cl_2(g) + 2e^- \rightarrow 2Cl^- \qquad E° = +1.36$$

The magnesium is oxidized. The equation for its half-reaction must be recorded as the reverse of the one in the table, including a change in the sign of $E°$:

$$Mg(s) \rightarrow Mg^{2+} + 2e^- \qquad E° = +2.37$$

(Moving the $2e^-$ to the right of the arrow permits us to show the loss of electrons.) The two half-reactions are then combined:

reduction: $Cl_2(g) + 2e^- \rightarrow 2Cl^-$ $E° = +1.36$
oxidation: $Mg(s) \rightarrow Mg^{2+} + 2e^-$ $E° = +2.37$

$$Mg(s) + Cl_2(g) \rightarrow Mg^{2+} + 2Cl^- \qquad E° = +3.73$$

Sometimes it is necessary to multiply one or both of the half-reactions by a coefficient in order to balance the electron transfer. Consider, for example, the reaction between sodium and chlorine to produce sodium chloride:

$$2Na + Cl_2 \rightarrow 2NaCl$$

The two half-reactions for this reaction are

$$Na° \rightarrow Na^+ + e^- \qquad E° = +2.71$$
$$Cl_2° + 2e^- \rightarrow 2Cl^- \qquad E° = +1.36$$

In order to balance the electron transfer, the sodium half-reaction must be multiplied by 2:

$2Na° \rightarrow 2Na^+ + 2e^-$ $E° = +2.71$ (Note: $E°$ is not multiplied)
$Cl_2° + 2e^- \rightarrow 2Cl^-$ $E° = +1.36$

$$2Na° + Cl_2° \rightarrow 2Na^+ + 2Cl^- \text{ (or 2NaCl)} \qquad E° = +4.07$$

When combining half-reactions, in order for the reaction to be spontaneous, the resultant $E°$ must be positive. Metals with negative reduction potentials (positive oxidation potentials) will react spontaneously with acids to produce hydrogen gas. For example, if magnesium metal were added to an acid solution, the half-reactions would be:

$$Mg° \rightarrow Mg^{2+} + 2e^- \qquad E° = +2.37$$
$$\underline{2H^+ + 2e^- \rightarrow H_2° \qquad E° = \ \ 0.00}$$
$$Mg° + 2H^+ \rightarrow Mg^{2+} + H_2° \qquad E° = +2.37$$

The positive value of $E°$ assures us that the reaction will occur spontaneously. Metals with positive reduction potentials (negative oxidation potentials) will not liberate hydrogen from acid solutions because the reaction will not occur at ordinary conditions. For example, if copper were added to hydrochloric acid, the half-reactions for any possible reaction would be:

$$Cu° \rightarrow Cu^{2+} + 2e^- \qquad E° = -0.34$$
$$\underline{2H^+ + 2e^- \rightarrow H_2° \qquad E° = \ \ 0.00}$$
$$E° = -0.34$$

The negative value of $E°$ tells us that no reaction will occur between copper and hydrochloric acid.

Equilibrium

The $E°$ values provided by tables of standard electrode potentials are for definite concentrations. As a reaction proceeds, these concentrations change and the measured voltage diminishes as equilibrium is approached. When equilibrium is reached, the voltage is equal to zero and the cell is labeled as a "dead" cell.

Electrodes

Cathode. The name *cathode* always applies for the electrode at which *reduction* occurs. In electrochemical cells this is the positive electrode. In an electrolytic cell the cathode is the negative electrode.

Anode. The name *anode* always applies for the electrode at which *oxidation* occurs. In an electrolytic cell the anode is the positive electrode and in an electrochemical cell, the anode is the negative electrode.

SAMPLE PROBLEM

Given the half-cells Cu°/Cu²⁺ (1.0M) and Zn°/Zn²⁺ (1.0M) and the diagram of the two half cells

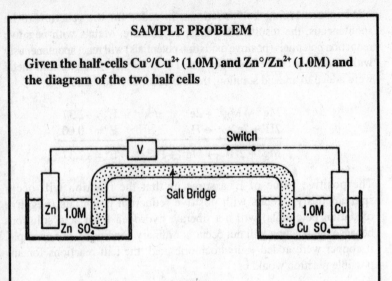

Figure 8-3.

complete the following:

1. oxidation half reaction =
   reduction half-reaction =
2. Label the anode and cathode
3. Show direction of electron flow
4. show direction of ion flow
5. Calculate E° for the cell

Solution

1. The table shows that the copper ion is more easily reduced. Therefore, the copper is the cathode, the zinc is the anode.
2. oxidation: $Zn° \rightarrow Zn^{2+} + 2e^-$ $E° = +0.76$
   reduction: $Cu^{2+} + 2e^- \rightarrow Cu°$ $E° = +0.34$
3. Electrons will flow from the anode (Zn) through the wire to the cathode (Cu).
4. In the salt bridge, positive ions flow toward the cathode and negative ions flow toward the anode.
5. E° for the cell: + .34
                            +0.76
                           ‾‾‾‾‾‾‾
                            1.10 v

ELECTROPLATING

Electroplating is the electrochemical process by which one metal is plated onto another. In this process, an electric current is used to produce a chemical reaction. The material to be plated is the cathode; the metal used for plating is the anode. The electrolyte used is a salt of the anode. The diagram below shows the electroplating of a tin spoon with silver.

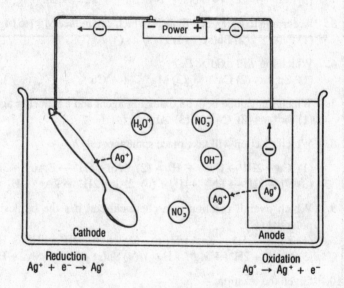

Reduction	Oxidation
$Ag^+ + e^- \rightarrow Ag^\circ$	$Ag^\circ \rightarrow Ag^+ + e^-$

Figure 8-4. Electroplating silver.

QUESTIONS

1. Given the reaction:

$$Ca(s) + Cu^{2+}(aq) \rightarrow Ca^{2+}(aq) + Cu(s)$$

What is the correct reduction half-reaction?
(1) $Cu^{2+}(aq) + 2e^- \rightarrow Cu(s)$ (2) $Cu^{2+}(aq) \rightarrow Cu(s) + 2e^-$
(3) $Cu(s) + 2e^- \rightarrow Cu^{2+}(aq)$ (4) $Cu(s) \rightarrow Cu^{2+}(aq) + 2e^-$

2. In the reaction $Mg + Cl_2 \rightarrow MgCl_2$, the correct half-reaction for the oxidation that occurs is
(1) $Mg + 2e^- \rightarrow Mg^{2+}$ (2) $Cl_2 + 2e^- \rightarrow 2Cl^-$
(3) $Mg \rightarrow Mg^{2+} + 2e^-$ (4) $Cl_2 \rightarrow 2Cl^- + 2e^-$

3. Based on Reference Table N, which of the following ions is most easily oxidized? (1) F^- (2) Cl^- (3) Br^- (4) I^-

4. Based on Reference Table N, which half-cell has a greater reduction potential than the standard hydrogen half-cell?

(1) $Na^+ + e^- \rightarrow Na(s)$ (2) $Ni^{2+} + 2e^- \rightarrow Ni(s)$
(3) $Pb^{2+} + 2e^- \rightarrow Pb(s)$ (4) $Sn^{4+} + 2e^- \rightarrow Sn^{2+}$

5. According to Reference Table N, which will reduce Mg^{2+} to $Mg(s)$?
(1) $Fe(s)$ (2) $Ba(s)$ (3) $Pb(s)$ (4) $Ag(s)$

6. Which ion will oxidize Fe?
(1) Zn^{2+} (2) Ca^{2+} (3) Mg^{2+} (4) Cu^{2+}

7. Which ion can be both an oxidizing agent and a reducing agent?
(1) Sn^{2+} (2) Cu^{2+} (3) Al^{3+} (4) Fe^{3+}

8. Which reaction will take place spontaneously?

(1) $Cu + 2H^+ \rightarrow Cu^{2+} + H_2$ (2) $2Au + 6H^+ \rightarrow 2Au^{3+} + 3H_2$
(3) $Pb + 2H^+ \rightarrow Pb^{2+} + H_2$ (4) $2Ag + 2H^+ \rightarrow 2Ag^+ + H_2$

9. Which overall reaction in a chemical cell has the highest net potential (E^0)?
(1) $Zn(s) + 2H^+ \rightarrow Zn^{2+} + H_2(g)$ (2) $Ni(s) + 2H^+ \rightarrow Ni^{2+} + H_2(g)$
(3) $Mg(s) + 2H^+ \rightarrow Mg^{2+} + H_2(g)$ (4) $Sn(s) + 2H^+ \rightarrow Sn^{2+} + H_2(g)$

10. Given the reaction:

$$2Al(s) + 3Pb^{2+}(aq) \rightarrow 2Al^{3+}(aq) + 3Pb(s)$$

The potential for (E^0) for the overall reaction is (1) 1.53 V
(2) 1.79 V   (3) 2.93 V   (4) 3.71 V

Base your answers to questions 11 and 12 on the following reaction:
$$Mg(s) + 2Ag^+(aq) \rightarrow Mg^{2+}(aq) + 2Ag(s)$$

11. Which species undergoes a loss of electrons?
(1) $Mg(s)$ (2) $Ag^+(aq)$ (3) $Mg^{2+}(aq)$ (4) $Ag(s)$

12. What is the cell voltage (E^0) for the overall reaction?

    (1) + 1.57 V   (2) +2.37 V   (3) +3.17 V   (4) +3.97 V

13. Given the chemical cell reaction:

$$2Ag^+ + Zn^0 \rightarrow 2Ag^0 + Zn^{2+}$$

What is the net potential (E^0) for the cell?

    (1) 1.56 V   (2) 2.36 V   (3) 0.84 V   (4) 0.04 V

Base your answers to questions 14 and 15 on the diagram of the chemical cell shown below. The reaction occurs at 1 atmosphere and 298 K.

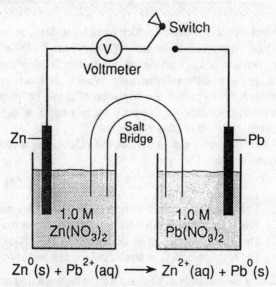

$$Zn^0(s) + Pb^{2+}(aq) \longrightarrow Zn^{2+}(aq) + Pb^0(s)$$

14. When the switch is closed, what occurs?

    (1) Pb is oxidized and electrons flow to the Zn electrode.

    (2) Pb is reduced and electrons flow to the Zn electrode.

    (3) Zn is oxidized and electrons flow to the Pb electrode.

    (4) Zn is reduced and electrons flow to the Pb electrode.

15. When the switch is closed, the cell voltage (E^0) is

    (1) +0.63 V   (2) +0.89 V   (3) –0.63 V   (4) –0.89 V

16. The diagram below represents an electrochemical cell

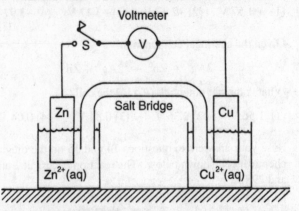

When switch S is closed, which particles undergo reduction?
(1) Zn^{2+} ions   (2) Zn atoms   (3) Cu^{2+} ions   (4) Cu atom

17. A chemical cell is made up of two half-cells connected by a salt bridge and an external conductor. What is the function of the salt bridge? (1) to permit the migration of ions (2) to prevent the migration of ions (3) to permit the mixing of solutions (4) to prevent the flow of electrons

18. Which statement best describes the reaction represented by the equation below?

$$2NaCl + 2H_2O + electricity \rightarrow Cl_2 + H_2 + 2NaOH$$

(1) The reaction occurs in a chemical cell and releases energy.
(2) The reaction occurs in a chemical cell and absorbs energy.
(3) The reaction occurs in an electrolytic cell and releases energy.
(4) The reaction occurs in an electrolytic cell and absorbs energy.

19. During the electrolysis of fused KBr, which reaction occurs at the positive electrode? (1) Br^- ions are oxidized. (2) Br^- ions are reduced. (3) K^+ ions are reduced. (4) K^+ ions are oxidized.

20. In an electrolytic cell, a Cl^- ion would be attracted to the (1) positive electrode and oxidized (2) positive electrode and reduced (3) negative electrode and oxidized (4) negative electrode and reduced

21. What occurs when an electrolytic cell is used for silverplating a spoon? (1) A chemical reaction produces an electrical current. (2) An electric current produces a chemical reaction. (3) An oxidation reaction takes place at the cathode. (4) A reduction reaction takes place at the anode.

22. The type of reaction in an electrochemical cell is best described as a (1) spontaneous oxidation reaction, only (2) nonspontaneous oxidation reaction, only (3) spontaneous oxidation-reduction reaction (4) nonspontaneous oxidation-reduction reaction

Base your answers to questions 23 and 24 on the equation and diagram below which represents an electrochemical cell at 298 K and 1 atmosphere.

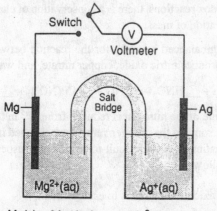

$$Mg(s) + 2Ag^+(aq) \longrightarrow Mg^{2+}(aq) + 2Ag(s)$$

23. Which species is oxidized when the switch is closed? (1) Mg(s) (2) Mg^{2+} (aq) (3) Ag(s) (4) Ag^+(aq)

24. When the switch is closed, electrons flow from
    (1) Mg(s) to Ag(s)   (2) Ag (s) to Mg(s)
    (3) Mg^{2+} (aq) to Ag^+ (aq)   (4) Ag^+ (aq) to Mg^{2+} (aq)

25. Given the reaction:

$$Fe + Sn^{2+} \rightarrow Fe^{2+} + Sn$$

    What is the potential difference (E^0) of this cell?
    (1) 0.14 V   (2) 0.31 V   (3) 0.45 V   (4) 0.59 V

26. In both the electrochemical cell and the electrolytic cell, the anode is the electrode at which (1) reduction occurs and electrons are lost (2) reduction occurs and protons are lost (3) oxidation occurs and electrons are lost (4) oxidation occurs and protons are lost

27. Based on Reference Table N, which reaction will take place spontaneously?
    (1) $Mg(s) + Ca^{2+}$ (aq) $\rightarrow Mg^{2+}$(aq) + Ca(s)
    (2) $Ba(s) + 2Na^+$(aq) $\rightarrow Ba^{2+}$(aq) + 2Na(s)
    (3) $Cl_2(g) + 2F^-$ (aq) $\rightarrow 2Cl^-$ (aq) + $F_2(g)$
    (4) $I_2(g) + 2Br^-$ (aq) $\rightarrow 2I^-$ (aq) + $Br_2(g)$

BALANCING REDOX REACTIONS

Two fundamental principles related to redox reactions provide the basis for a structure to be used in balancing equations for redox reactions. These principles are:

1. In all redox reactions, the electrons lost must be equal to the electrons gained.
2. In all redox reactions, there is a conservation of charge as well as a conservation of mass.

Consider the unbalanced equation for the reaction between copper and nitric acid to produce nitric oxide, copper nitrate, and water:

$$\_\_Cu + \_\_HNO_3 \rightarrow \_\_NO + \_\_Cu(NO_3)_2 + \_\_H_2O$$

Notice that some of the nitrogen is reduced from N^{+5} in the HNO_3 to N^{+2} in the NO. But some of the nitrogen remains unchanged in the $Cu(NO_3)_2$. Thus, this equation is very difficult to balance by inspection. It can be balanced as follows:

1. Assign oxidation numbers to each element.

$$Cu^\circ + H^{+1}N^{+5}O_3{}^{-2} \rightarrow N^{+2}O^{-2} + Cu^{+2}(N^{+5}O_3{}^{-2})_2 + H_2{}^{+1}O^{-2}$$

2. Determine the oxidation and reduction reactions.

   oxidation: $Cu^\circ \rightarrow Cu^{+2}$
   reduction: $N^{+5} \rightarrow N^{+2}$

3. Write the equations.

   $Cu^\circ - 2e^- \rightarrow Cu^{+2}$
   $N^{+5} + 3e^- \rightarrow N^{+2}$

4. When necessary, multiply each half reaction by the other half reaction's number of e^- in order to balance electrons gained and lost. Then combine the balanced partial equations.

$$3\,(Cu^\circ \quad \rightarrow Cu^{2+} + 2e^-) = \quad 3Cu^\circ \quad \rightarrow 3Cu^{2+} + 6e^-$$
$$2\,(N^{+5} + 3e^- \rightarrow N^{+2}) \quad = \quad \underline{2N^{+5} + 6e^- \rightarrow 2N^{+5}}$$
$$3Cu^\circ + 2N^{+5} \rightarrow 3Cu^{+2} + 2N^{+2}$$

This type of equation is known as a **net equation.** It does not include spectator ions or atoms that are not involved in oxidation or reduction.

5. Add the coefficients from Step 4 into the original unbalanced equation.

$$3Cu + 2HNO_3 \rightarrow 2NO + 3Cu(NO_3)_2 + H_2O$$

6. Complete the task of balancing by inspection in order to satisfy conservation of matter.

$$3Cu + 8HNO_3 \rightarrow 2NO + 3Cu(NO_3)_2 + 4H_2O$$

QUESTIONS

1. Given the reaction:

$$\_Hg^{2+} + \_Ag \rightarrow \_Hg + \_Ag^+$$

When the equation is completely balanced using the smallest whole-number coefficients, the coefficient of Hg will be
(1) 1   (2) 2   (3) 3   (4) 4

2. Given the unbalanced equation which represents aluminum metal reacting with an acid:

$$Al + H^+ \rightarrow Al^{+3} + H_2$$

What is the total number of moles of electrons lost by 1 mole of aluminum?   (1) 6   (2) 2   (3) 3   (4) 13

3. Given the unbalanced equation:

$$\_Fe + \_Ag^+ \rightarrow \_Ag + \_Fe^{3+}$$

When the equation is correctly balanced using smallest whole numbers, the coefficient of Ag+ is   (1) 5   (2) 2   (3) 3   (4) 4

4. Given the equation:

$$3Cu + 8HNO_3 \rightarrow 3Cu(NO_3)_2 + 2NO + 4H_2O$$

What is the total number of moles of electrons lost by the copper as it completely reacts with 8 moles of nitric acid?   (1) 1   (2) 6   (3) 8   (4) 4

5. How many moles of electrons are needed to reduce one mole of Cu^{2+} to Cu^{1+}?   (1) 1   (2) 2   (3) 3   (4) 4

6. The following equation represents the reaction for a zinc-copper chemical cell:

$$Zn(s) + Cu^{2+}(aq) \rightarrow Zn^{2+}(aq) + Cu(s).$$

If 0.1 mole of copper is deposited on the copper electrode, the mass of the zinc electrode will (1) decrease by 6.5 g (2) increase by 6.5 g (3) decrease by 13 g (4) increase by 13 g

7. How many moles of electrons would be required to completely reduce 1.5 moles of Al^{3+} to Al? (1) 0.50 (2) 1.5 (3) 3.0 (4) 4.5

8. When the equation

$$HNO_3 + MnCl_2 + HCl \rightarrow NO + MnCl_4 + H_2O$$

is completely balanced using the *smallest* whole number coefficients, the coefficient of the HNO_3 will be (1) 5 (2) 2 (3) 3 (4) 6

9. When the equation

$$\_\_\_Cu + \_\_\_HNO_3 \rightarrow \_\_\_Cu(NO_3)_2 + 2NO + \_\_\_H_2O$$

is completely balanced using whole numbers, the coefficient of the HNO_3 will be (1) 8 (2) 2 (3) 6 (4) 4

10. Given the unbalanced equation:

$$Ca^0 + Al^{3+} \rightarrow Ca^{2+} + Al^0$$

When the equation is completely balanced with the smallest whole-number coefficients, what is the coefficient of Ca^0?
(1) 1 (2) 2 (3) 3 (4) 4

UNIT 9. ORGANIC CHEMISTRY

Organic chemistry is the name given to the study of carbon and carbon compounds. Carbon atoms have a very strong tendency to form four covalent bonds. Because of this tendency, carbon atoms produce an enormous variety of compounds. They form bonds not only with atoms of other elements, they also form bonds with other carbon atoms. The number of carbon compounds (or organic compounds) is far greater than the number of inorganic compounds-perhaps as much as thirty times greater.

Living things contain many carbon compounds. Products of living things, such as petroleum, wood, and coal supply the raw materials from which most organic chemicals are obtained.

CHARACTERISTICS OF ORGANIC COMPOUNDS

Organic compounds are subject to all the chemical principles that apply to inorganic compounds. One of the major differences between organic and inorganic compounds is the type of bonding present. Bonding in inorganic compounds includes ionic, covalent, and coordinate covalent bonds. Most organic compounds result exclusively from covalent bonds—pairs of shared electrons.

Some important characteristics include the following:

1. Organic compounds are generally nonpolar.
2. Only a few organic compounds will dissolve in water. These include acetic acid, various sugars, and certain alcohols. Many organic compounds are soluble in nonpolar solvents. These solvents are usually organic compounds themselves.
3. Most organic compounds are nonelectrolytes. Organic acids are exceptions—they are weak electrolytes.
4. Organic compounds have low melting points, due to the fact that the compounds are held together by weak intermolecular forces.
5. Reaction rates of organic compounds are slower than those of inorganic compounds. In contrast to their weak intermolecular forces, the covalent bonds within the organic molecules are very strong. Activation energy, therefore, is very high, and catalysts are often used to increase reaction rates.

Bonding

Carbon atoms generally form compounds by covalent bonding. Hybridization of s and p orbitals provides carbon atoms with electron configurations that consist of four bonding orbitals. The four "valence" electrons seem to produce four single bonds that are so related to each other in space as to form regular tetrahedrons.

When carbon atoms form bonds with other carbon atoms, they can share one, two, or even three pairs of electrons, the latter cases resulting in double and triple bonds. Because the bonding is covalent, carbon compounds are molecular.

Structural Formulas

The importance of the three-dimensional nature of organic compounds has led to the widespread use of models to illustrate the bonding within organic molecules. Another device for representing the arrangement of atoms in an organic compound is the **structural formula.** In a structural formula, each covalent bond is represented by a short, straight line between atoms. Consider the structural formula of methane, CH_4:

Methane

This structural formula provides an approximation of the shape of the molecule. A ball-and-stick model can be used to show, in three dimensions, the actual tetrahedral arrangement of the bonds in methane.

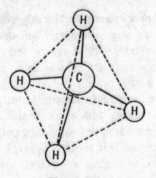

Figure 9-1.

Isomers

Many organic compounds have the same molecular formula—that is, the same atoms in the same molar ratios—but display vastly different properties. Investigations have revealed that while molecular formulas may be the same, the arrangements of the atoms are quite different. Compounds that have the same molecular formula but different structural formulas are called **isomers.** Acetone and propanal are examples of isomers. Both have the molecular formula C_3H_6O. Structural formulas of these two compounds show the different arrangements of the atoms.

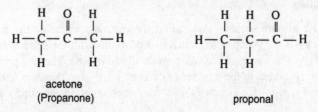

acetone
(Propanone)   proponal

Structural formulas are somewhat cumbersome and require considerable space. An intermediate device, between molecular and structural formulas, is often used to represent organic compounds on a single type line. These formulas show the atoms in the "order" in which they are arranged in the molecule. For example, the formula for acetone is written CH_3COCH_3; that for propanal is written CH_3CH_2CHO.

Saturated and Unsaturated Compounds

Saturated compounds are organic compounds in which all the carbon-to-carbon bonds are formed by the sharing of single pairs of electrons. In other words, all carbon-to-carbon bonds are single bonds.

Unsaturated compounds are those in which at least one of the carbon-to-carbon bonds is a double or triple bond. A double bond is formed from two shared pairs of electrons. A triple bond contains three shared pairs of electrons.

HOMOLOGOUS SERIES OF HYDROCARBONS

Compounds containing only hydrogen and carbon atoms are called **hydrocarbons.** Many organic compounds can be thought of as being related to, or derived from, hydrocarbons.

The International Union of Pure and Applied Chemistry (IUPAC) has devised a system for naming organic compounds. The system was devised to eliminate the confusion caused by the use of common names. In many cases, the IUPAC names have been derived from the hydrocarbons to which the compounds are related.

A **homologous series** is a group of organic compounds with similar properties and related structures. The formulas of the members of a homologous series differ from each other by some common increment. As molecular size increases among members of a homologous group, so too does the relative strength of the van der Waals forces holding the molecules together. The increase in van der Waals forces results in increased boiling points and freezing (melting) points. Compounds with low molecular masses tend to be gases. Those with the highest molecular masses tend to be solids under normal conditions of temperature and pressure.

Alkanes

The **alkane** series—also called the methane series or the paraffin series—is the group of saturated hydrocarbons with the general formula C_nH_{2n+2}. The names of the compounds all end in -*ane*. The first part of the name is related to the number of carbons: *meth-* for 1 carbon, *eth-* for 2 carbons, *prop-* for 3 carbons, *but-* for 4 carbons and *pent-* for 5 carbons.

TABLE 9-1. FIRST FIVE MEMBERS OF THE ALKANE SERIES

Hydrocarbon	Molecular formula	Structural formula
methane	CH_4	H | H—C—H | H
ethane	C_2H_6	H H | | H—C—C—H | | H H
propane	C_3H_8	H H H | | | H—C—C—C—H | | | H H H
butane	C_4H_{10}	H H H H | | | | H—C—C—C—C—H | | | | H H H H
pentane	C_5H_{12}	H H H H H | | | | | H—C—C—C—C—C—H | | | | | H H H H H

The alkane series begin to show isomerism with butane, C_4H_{10}.

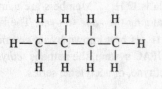

normal or n-butane

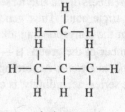

isobutane
(methyl propane)

Alkenes

The members of the **alkene** series are unsaturated hydrocarbons containing one double bond. Alkenes have the general formula C_nH_{2n} The alkenes are named by changing the ending of the corresponding alkane from -*ane* to -*ene* The first four alkenes are ethene, propene, butene, and pentene. The alkene series is also called the ethylene or the olefin series. It should be noted here that unsaturated hydrocarbons with more than one double bond are *not* considered members of the alkene series. The **dienes**, for example, is the group whose members each have two double bonds.

TABLE 9-2. FIRST FOUR MEMBERS OF THE ALKENE SERIES

Hydrocarbon	Molecular formula	Structural formula
ethene	C_2H_4	$\begin{array}{c} H \\ \diagdown \\ C=C \\ \diagup \\ H \end{array} \begin{array}{c} H \\ \diagup \\ \\ \diagdown \\ H \end{array}$
propene	C_3H_6	H—C=C—C—H (with H atoms)
butene	C_4H_8	H—C=C—C—C—H (with H atoms)
pentene	C_5H_{10}	H—C—C=C—C—C—H (with H atoms)

Alkynes

Members of the **alkyne** series are unsaturated hydrocarbons containing one triple bond. Their general formula is C_nH_{2n-2}. Members are named from the corresponding alkane by changing the *–ane* to *–yne*. The first member of the group $H-C \equiv C-H$ is best known by its common name, *acetylene*. According to the IUPAC system, the name is *ethyne*. The series, accordingly, is called the ethyne, or acetylene, series.

TABLE 9-3. FIRST FOUR MEMBERS OF THE ALKYNE SERIES

Hydrocarbon	Molecular formula	Structural formula
ethyne	C_2H_2	$H-C \equiv C-H$
propyne	C_3H_4	$H-C \equiv C-\underset{\underset{H}{\vert}}{\overset{\overset{H}{\vert}}{C}}-H$
2-butyne	C_4H_6	$H-\underset{\underset{H}{\vert}}{\overset{\overset{H}{\vert}}{C}}-C \equiv C-\underset{\underset{H}{\vert}}{\overset{\overset{H}{\vert}}{C}}-H$
2-pentyne	C_5H_8	$H-\underset{\underset{H}{\vert}}{\overset{\overset{H}{\vert}}{C}}-C \equiv C-\underset{\underset{H}{\vert}}{\overset{\overset{H}{\vert}}{C}}-\underset{\underset{H}{\vert}}{\overset{\overset{H}{\vert}}{C}}-H$

Benzene Series (Arenes)

Members of the **benzene** series have the general formula C_nH_{2n-6}. A significant difference from the alkanes, alkenes, and alkynes is that the compounds of the benzene series consist of closed chains, while those of the other three series are open chains. The benzene series is a series of **cyclic hydrocarbons** in which the simplest member is *benzene* C_6H_6. Toluene, C_7H_8, is the second member.

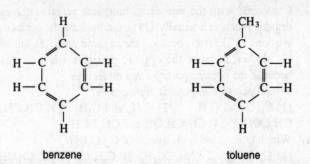

benzene   toluene

Actually, all the carbon-to-carbon bonds in the benzene ring are the same. They have structures and related properties intermediate between single and double bonds. To show this structure, the structural formula is better written:

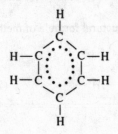

For the sake of simplicity chemists use

benzene

The members of the benzene series are also referred to as aromatic hydrocarbons.

QUESTIONS

1. A general characteristic of organic compounds is that they all (1) react vigorously (2) dissolve in water (3) are strong electrolytes (4) melt at relatively low temperatures

2. All organic compounds must contain the element (1) hydrogen (2) nitrogen (3) carbon (4) oxygen

3. The four single bonds of a carbon atom are spatially directed toward the corners of a regular (1) triangle (2) rectangle (3) square (4) tetrahedron

4. Compared with the rate of an inorganic reaction the rate of an organic reaction is usually (1) faster, because the organic particles are ions (2) faster, because the organic particles are molecules (3) slower, because the organic particles are ionic (4) slower, because the organic particles are molecules

5. Which pair of compounds are isomers?
(1) C_6H_6 and C_6H_{12} (2) C_2H_4 and C_2H_6 (3) CH_3CH_2OH and CH_3COOH (4) CH_3CH_2OH and CH_3OCH_3

6. Which compound is an isomer of C_4H_9OH?
(1) $C_3H_7CH_3$ (2) $C_2H_5OC_2H_5$ (3) $C_2H_5COOC_2H_5$ (4) CH_3COOH

7. As the number of carbon atoms in a hydrocarbon molecule increases, the number of possible isomers generally (1) decreases (2) increases (3) remains the same

8. In the alkane series, each molecule contains (1) only one double bond (2) two double bonds (3) one triple bond (4) all single bonds

9. Which is the structural formula of methane?

10. Which structural formula represents a saturated hydrocarbon?

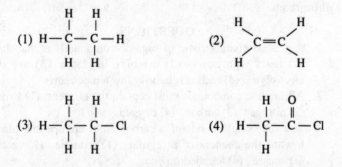

11. Which is the structural formula of ethene?

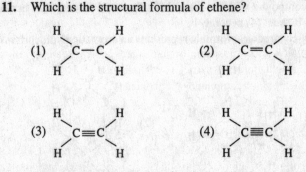

(1) (2) (3) (4)

12. What is the formula of pentene? (1) C_4H_8 (2) C_4H_{10} (3) C_5H_{10}
 (4) C_5H_{12}

13. What is the total number of pairs of electrons that one carbon atom shares with the other carbon atom in the molecule C_2H_4?
 (1) 1 (2) 2 (3) 3 (4) 4

14. Which formula represents an alkene? (1) CH_4 (2) C_2H_2
 (3) C_3H_6 (4) C_4H_{10}

15. In which hydrocarbon series does each molecule contain one triple bond? (1) alkane (2) alkene (3) alkyne (4) benzene

16. Which set of formulas represents members of the same homologous series? (1) C, CH_4, CH_4O (2) C_2H_4, C_3H_6, C_4H_8
 (3) C_2H_2, C_2H_4, C_2H_6 (4) CH_2, CH_3, CH_4

17. Which hydrocarbon is a member of the series with the general formula C_nH_{2n-2}? (1) ethyne (2) ethene (3) butane (4) benzene

18. Given the compound:

$$H-\overset{\overset{\displaystyle H}{|}}{\underset{\underset{\displaystyle H}{|}}{C}}-\overset{\overset{\displaystyle H}{|}}{C}=\overset{\overset{\displaystyle H}{|}}{C}-\overset{\overset{\displaystyle H}{|}}{\underset{\underset{\displaystyle H}{|}}{C}}-H$$

What is the general formula of the hydrocarbon series of which this compound is a member? (1) C_nH_{2n+2} (2) C_nH_{2n} (3) C_nH_{2n-2}
 (4) C_nH_{2n-6}

19. Given the compounds:

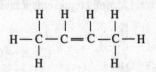

These compounds are both (1) alkynes (2) alkenes (3) isomers of butane (4) isomers of propane

20. A compound with the formula C_6H_6 is (1) toluene (2) benzene (3) butene (4) pentene

21. Which structural formula represents an aromatic hydrocarbon?

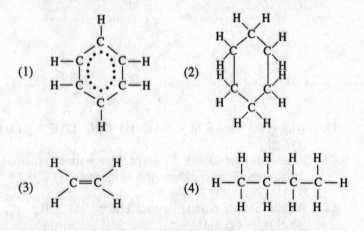

22. The compound $CH_3CH_2CH_2CH_3$ belongs to the series that has the general formula (1) C_nH_{2n-2} (2) C_nH_{2n+2} (3) C_nH_{n-6} (4) C_nH_{n+6}

23. Which diagram may be used to represent a benzene ring?

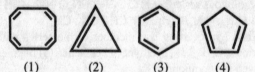

(1)   (2)   (3)   (4)

24. Which structural formula represents a saturated compound?

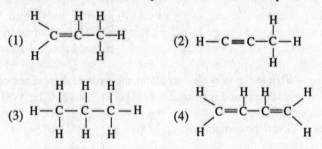

25. Which kind of bond is most common in organic compounds?
(1) covalent (2) ionic (3) hydrogen (4) electrovalent

26. Which is an isomer of $CH_3CH_2CH_2COOH$?
(1) $CH_3CH_2OCH_2CH_3$ (2) $CH_3CH_2CH_2OCH_3$
(3) $CH_3CH_2CH_2CH_2OH$ (4) $CH_3COOCH_2CH_3$

27. Which structural formula represents a compound that is an isomer of

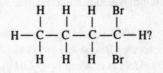

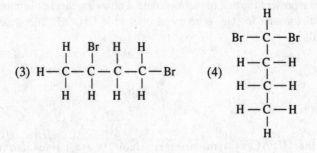

Other Organic Compounds

Other homologous series of organic compounds are formed by the replacement of one or more hydrogen atoms of a hydrocarbon by atoms of other elements. Members of these series are named from their corresponding hydrocarbons. However, they are not necessarily prepared directly from the compounds from which their names have been derived. Many of these groups have been classified according to the presence of some common particular arrangement of atoms known as **functional groups**. These groups provide characteristic properties to the compounds that contain them.

Alcohols

The functional group for this class of organic compounds is the **hydroxyl** (–OH) group. In alcohols, one or more hydrogens of a hydrocarbon have been replaced by this –OH group. Under ordinary conditions, no more than one –OH group can be attached to a single carbon atom. It should be noted that the –OH group of alcohols does *not* form the hydroxide ion in aqueous solutions. Therefore, alcohols are *not* bases.

Alcohols can be classified according to the *number* of –OH groups contained in each molecule. *Monohydroxy* (ℓ) alcohols contain *one* –OH group. *Dihydroxy* (ℓ) alcohols contain *two* –OH groups. Those alcohols containing *three* –OH groups are known as *trihydroxy* (ℓ) alcohols.

Monohydroxy Alcohols.

Primary Alcohols. Alcohols with one –OH attached to the end carbon of a hydrocarbon are called **primary** alcohols.

Since the functional group can be attached to any hydrocarbon, it is customary, when describing the classes of compounds to use the letter "R" to represent the rest of the molecule. Following this convention, the general formula for any primary alcohol is R–CH$_2$OH. The structural formula is

In the IUPAC system, primary alcohols are named from the corresponding hydrocarbon by replacing the final -*e* with -*ol*. Methanol, CH$_3$OH, the simplest primary alcohol, is formed by replacing one hydrogen of methane with an –OH group. Ethanol, C$_2$H$_5$OH, is formed by replacing one hydrogen of ethane with an –OH group. The structural formulas of these two alcohols are

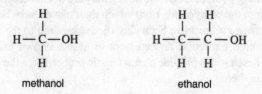

methanol   ethanol

Many alcohols are still called by their common names. These names were derived from a system in which the -*ane* of the hydrocarbon was replaced with -*yl*. Thus, methanol and ethanol are known, respectively, as methyl alcohol and ethyl alcohol.

TABLE 9-4. SEVERAL PRIMARY ALCOHOLS

Molecular formula	IUPAC name	Common name
CH_3OH	methanol	methyl alcohol
C_2H_5OH	ethanol	ethyl alcohol
C_3H_7OH	propanol	propyl alcohol
C_4H_9OH	butanol	butyl alcohol

Secondary alcohols. In secondary alcohols, the –OH group is attached to a carbon atom that is attached to two other carbon atoms. Secondary alcohols can be represented as

$$R_1 \!-\! \underset{\underset{H}{|}}{\overset{\overset{OH}{|}}{C}} \!-\! R_2$$

where R_1 and R_2 represent hydrocarbon radicals that may be the same or different. One example of a secondary alcohol is 2-propanol, or isopropanol. Its structural formula is

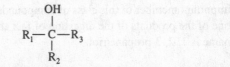

Tertiary alcohols. In tertiary alcohols, the –OH group is attached to a carbon atom that is attached to *three* other carbon atoms. Tertiary alcohols can be represented as

$$R_1 \!-\! \underset{\underset{R_2}{|}}{\overset{\overset{OH}{|}}{C}} \!-\! R_3$$

Since a minimum of four carbon atoms are needed to satisfy the definition, the simplest tertiary alcohol is tertiary butanol. Its structural formula is

```
        H   OH  H
        |   |   |
   H — C — C — C — H
        |   |   |
        H   |   H
        H — C — H
            |
            H
```

IUPAC has devised a system of numbering the carbon atoms in a hydrocarbon chain. The name of the compound is based on this numbering system. For example, tertiary butanol is named 2 methyl-2 propanol. The basic name comes from the longest carbon chain in the compound. In this case, the longest chain has 3 carbon atoms. Hence the prefix propan-. The carbons in the chain are numbered. The name of each branch is preceded by the number of the carbon atom to which it is attached.

Dihydroxy Alcohols

The dihydroxy alcohols contain two –OH groups. These alcohols are also known as *glycols*. The best known dihydroxy alcohol is probably ethylene glycol, the active ingredient in most permanent anti-freeze products. Its IUPAC name is 1, 2 ethanediol. Notice that the suffix *-diol* is added to the full name of the corresponding hydrocarbon.

1,2-ethanediol
(ethylene glycol)

The dihydroxy alcohols are also referred to as the dihydric alcohols.

Trihydroxy Alcohols

The trihydroxy (trihydric) alcohols have three –OH groups. The most important member of this class of compounds is best known as *glycerol,* one of the products of the digestion of fats and oils (lipids). The IUPAC name is 1, 2, 3-propanetriol.

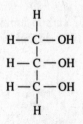

1,2,3-propanetriol
(glycerol)

ORGANIC ACIDS

The group of organic compounds known as the organic acids contain the functional group –COOH (carboxyl). Organic acids are named by replacing the final *-e* of the corresponding hydrocarbon with *-oic*, and adding the name acid. The first two members of this group are methanoic acid and ethanoic acid. These acids are better known by their common names, formic acid and acetic acid.

formic acid
(methanoic acid)

acetic acid
(ethanoic acid)

ALDEHYDES

Aldehydes contain the functional group $-\overset{\overset{\displaystyle H}{|}}{C}=O$. This is known as the **formyl** group. They are named by replacing the *-e* of the corresponding hydrocarbon with *-al*.

The primary alcohols can be oxidized to produce aldehydes.

$$2H-\underset{\underset{\displaystyle H}{|}}{\overset{\overset{\displaystyle H}{|}}{C}}-OH + O_2 \;\rightarrow\; 2H-\overset{\overset{\displaystyle H}{|}}{C}=O + 2H_2O$$

methanol   methanal

The common name for methanal is formaldehyde. Aldehydes are generally not very stable and are easily oxidized to organic acids.

KETONES

Ketones can be produced by the oxidation of secondary alcohols. The general formula for ketones is

$$R_1-\overset{\overset{\displaystyle O}{\|}}{C}-R_2$$

The functional group is $-\overset{\overset{\displaystyle O}{\|}}{C}-$.   They are named by replacing the -e of the corresponding hydrocarbon with -one.

The simplest ketone is one in which both R_1 and R_2 are methyl (CH_3) groups.

propanone

Propanone

Propanone is also knówn as dimethyl ketone, and by its most common name, acetone. With all of the different names for the same compound, it becomes readily apparent why IUPAC has tried to develop a universally accepted system for naming compounds.

ETHERS

The general formula for the ethers is R_1- O- R_2. Ethers can be produced by dehydration synthesis of two primary alcohols:

$$R_1–OH + HO–R_2 \rightarrow R_1–O–R_2 + H_2O$$

The functional group is —O—.
Diethyl ether, $C_2H_5OC_2H_5$, is commonly used as a solvent and anesthetic.

QUESTIONS

1. Methanol is classified as a   (1) monohydroxy alcohol
    (2) secondary alcohol   (3) tertiary alcohol   (4) dihydroxy alcohol
2. Which compound is an electrolyte?
    (1) C_2H_5OH (2) $C_3H_5(OH)_3$ (3) CH_3OH (4) CH_3COOH

3. Which is the structural formula of a primary alcohol?

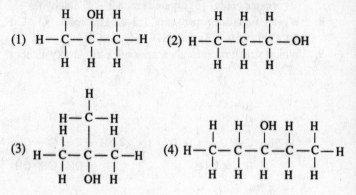

(1)

(2)

(3)

(4)

4. Which is the formula of methanal?

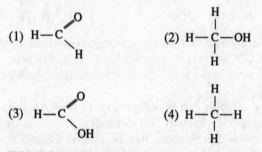

(1)

(2)

(3)

(4)

5. Which formula represents an organic acid?   (1) HCOOCH₃
(2) CH₃CH₂OH   (3) CH₃COCH₃   (4) HCOOH

6. Which compound is an isomer of propanone

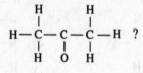 ?

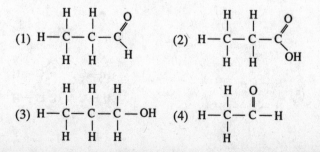

(1)

(2)

(3)

(4)

7. Ethers can be synthesized by dehydration of (1) primary alcohols
   (2) organic acids (3) organic ketones (4) aldehydes

8. Which formula represents 1,2-ethanediol? (1) $C_2H_4(OH)_2$
   (2) $C_3H_5(OH)_3$ (3) $Ca(OH)_2$ (4) $Co(OH)_3$

9. Which structural formula represents a trihydroxy alcohol?

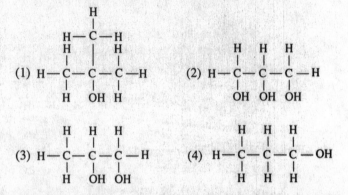

10. Which statement is true for a compound whose formula is
    CH_3CH_2COOH? (1) It is an alcohol. (2) It is an acid. (3) Its
    solution turns litmus blue. (4) Its solution turns phenolphthalein
    pink.

11. Primary alcohols can be dehydrated to produce (1) ethers
    (2) organic acid (3) esters (4) aldehydes

12. Which class of compounds has the general formula $R_1–O–R_2$?
    (1) esters (2) alcohols (3) ethers (4) aldehydes

13. Which structural formula represents a dihydroxy alcohol?

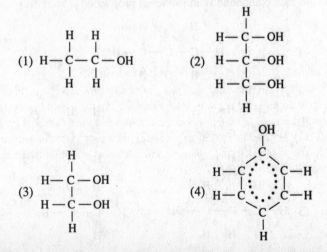

14. What is the minimum number of carbon atoms a ketone may contain? (1) 1 (2) 2 (3) 3 (4) 4

15. Which is the formula for methanoic acid?
    (1) CH_3OH (2) C_2H_5OH (3) HCOOH (4) $HC_2H_3O_2$

16. Which compound is a trihydroxy alcohol?
    (1) glycerol (2) butanol (3) ethanol (4) methanol

17. What is the total number of hydroxyl groups contained in one molecule of 1,2-ethanediol? (1) 1 (2) 2 (3) 3 (4) 4

18. Which structural formula represents an aldehyde?

19. Molecules of 1-propanol and 2-propanol have different
    (1) percentage compositions   (2) molecular masses
    (3) molecular formulas   (4) structural formulas

20. In an aqueous solution, which compound will be acidic?
    (1) CH_3COOH (2) CH_3CH_2OH (3) $C_3H_5(OH)_3$ (4) CH_3OH

ORGANIC REACTIONS

As noted earlier, organic reactions typically proceed at much slower rates than do inorganic reactions. For this reason, the use of catalysts is a common practice. In many organic reactions, only the functional group is involved. The greater part of the reacting molecules remain unchanged during the course of the reaction, and can easily be identified in the products.

Substitution

As the name implies, substitution reactions involve replacing one kind of atom or group with another kind of atom or group. For the saturated hydrocarbons, all substitution reactions (except for the special cases of combustion and thermal decomposition) involve replacement of hydrogen atoms. The halogen (F, Cl, Br, I) derivatives of the alkanes can be prepared by substitution reactions between the alkane and the halogen. The general term for these reactions is *halogen substitution*.

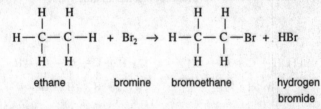

ethane   bromine   bromoethane   hydrogen
                                                    bromide

Preparation of the halogen derivatives by substitution always results in a by-product of the hydrogen halide.

Addition

Addition reactions involve *adding* two or more atoms to carbon atoms that are attached to other carbon atoms by double or triple bonds. Thus, addition reactions are generally limited to the unsaturated hydrocarbons. Addition reactions take place more easily than substitution reactions. Their rates are often as fast as those of ionic reactions. As a result, unsaturated compounds are considered more reactive than saturated compounds. Furthermore, those with triple bonds (alkynes) tend to be more reactive than those with double bonds (alkenes). Addition of hydrogen to an unsaturated compound, however, usually requires the presence of a catalyst and an elevated temperature. The hydrogen addition reaction is called *hydrogenation*. Addition reactions between unsaturated hydrocarbons and chlorine and bromine to produce halogen derivatives take place at room temperature.

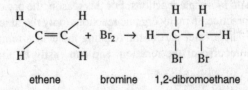

ethene   bromine   1,2-dibromoethane

Addition reactions are characterized by the formation of a single product. This is in contrast with substitution reactions in which more than one product is typical.

Fermentation

Fermentation is a process ordinarily associated with living systems. Enzymes produced by the living things serve as catalysts for the reactions in which organic molecules are broken down. For example, the fermentation of glucose is shown:

$$C_6H_{12}O_6 \xrightarrow[\text{(from yeast)}]{\text{zymase}} 2C_2H_5OH + 2CO_2$$

$$\text{glucose} \qquad\qquad\qquad \text{ethanol} \qquad \text{carbon dioxide}$$

Esterification

Esterification derives its name from the name of the products, esters. Esterification involves the reaction between an organic acid and an alcohol to produce an ester and water. Esters have a first and last name. The first name is derived from the alcohol name with a -*yl* ending. The last name comes from the organic acid with an -*ate* ending.

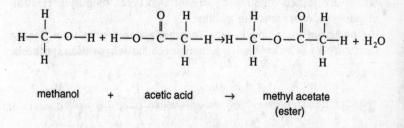

methanol   +   acetic acid   →   methyl acetate
(ester)

Since esterification involves an acid and an –OH group, it is often compared with neutralization of inorganic acids with bases, producing salts and water. Esterification, however, is *not* an ionic reaction, and esters are covalent compounds. Esterification is a slow reaction, usually requiring a catalyst, and it is reversible. Esters are responsible for the aromas associated with many fruits, flowers, and leaves. Lipids (fats and oils) are esters formed from esterification of glycerol (1, 2, 3 propantriol) by long-chain organic acids (fatty acids).

Saponification

The hydrolysis of fats by bases is called *saponification*. The organic salts that are produced are soaps. Glycerol is a second product of saponification reactions and is considered a byproduct in the manufacture of soap.

Oxidation

Saturated hydrocarbons react readily with oxygen under conditions of combustion. Such reactions result in the oxidation of the carbon to carbon monoxide or carbon dioxide, depending on the amount of oxygen available. The oxygen is reduced to water as well as the oxide of carbon. Oxidation reactions have great significance because of the liberation of energy associated with them. Energy is derived from fuels by combustion and from food by cellular respiration, both processes involving oxidation reactions.

Polymerization

Polymerization is a name given to reactions in which large molecules are made from smaller molecules. Polymerization occurs in nature in the production of proteins and starches by living organisms. Synthetic rubbers, plastics, and fibers are results of polymerization reactions.

Polymers are composed of many repeating units, called *monomers,* which are joined together by one of two types of polymerization reactions—condensation or addition.

Condensation. Condensation polymerization results from joining monomers by dehydration. It is sometimes called **dehydration synthesis.**

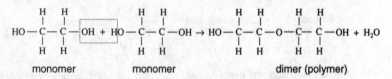

monomer   monomer   dimer (polymer)

This process may be repeated to produce a long-chain polymer. Monomers involved in condensation must have at least two functional groups. Examples of condensation polymers include silicones, polyesters, polyamides, phenolic plastics, and nylons.

Addition. Addition polymerization, as do all addition reactions, involves opening of double and triple bonds of unsaturated hydrocarbons.

Vinyl plastics, such as polyethylene and polystyrene, are examples of addition polymers.

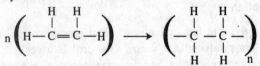

ethylene
(the monomer)

polyethylene
(the polymer)

QUESTIONS

1. The organic reaction
   $HCOOH + CH_3CH_2CH_2CH_2OH \rightarrow HCOOCH_2CH_2CH_2CH_3 + HOH$,
   is an example of   (1) fermentation   (2) esterification
   (3) polymerization   (4) saponification

2. Alkanes differ from alkenes in that alkanes (1) are hydrocarbons
   (2) are saturated compounds (3) have the general formula C_nH_{2n}
   (4) undergo addition reactions

3. Which molecule is represented by X in the reaction

$$H - C \equiv C - H + 2Br_2 \rightarrow X \, ?$$

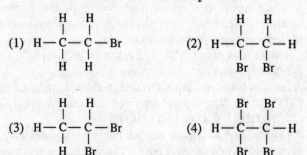

   (1) ... (2) ... (3) ... (4) ...

4. Which equation represents an esterification reaction?
   (1) $C_6H_{12}O_6 \rightarrow 2C_2H_5OH + 2CO_2$
   (2) $C_5H_{10} + H_2 \rightarrow C_5H_{12}$
   (3) $C_3H_8 + Cl_2 \rightarrow C_3H_7Cl + HCl$
   (4) $HCOOH + CH_3OH \rightarrow HCOOCH_3 + HOH$

5. What is the correct IUPAC name for

$$
\begin{array}{ccccc}
& H & Cl & Cl & \\
& | & | & | & \\
H - & C & - C & - C & - H \, ? \\
& | & | & | & \\
& H & H & H &
\end{array}
$$

   (1) 1,2-dichlorobutane   (2) 2,3-dichlorobutane
   (3) 1,2-dichloropropane   (4) 2,3-dichloropropane

6. Which equation represents a subsititution reaction?
   (1) $CH_4 + 2O_2 \rightarrow CO_2 + 2H_2O$
   (2) $C_2H_4 + Br_2 \rightarrow C_2H_4Br_2$
   (3) $C_3H_6 + H_2 \rightarrow C_3H_8$
   (4) $C_4H_{10} + Cl_2 \rightarrow C_4H_9Cl + HCl$

7. Which reaction produces ethyl alcohol as one of the principal
   products? (1) an esterification reaction (2) a neutralization
   reaction (3) a saponification reaction (4) a fermentation reaction

8. Which is the product of the reaction between ethene and chlorine?

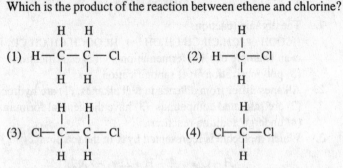

9. In which type of reaction are long-chain molecules formed from smaller molecules? (1) substitution (2) saponification (3) fermentation (4) polymerization

10. Cellulose is an example of (1) a synthetic polymer (2) a natural polymer (3) an ester (4) a ketone

11. A reaction between CH_3COOH and an alcohol produced the ester CH_3COOCH_3. The alcohol used in the reaction was (1) CH_3OH (2) C_2H_5OH (3) C_3H_7OH (4) C_4H_9OH

12. An alcohol and an organic acid are combined to form water and a compound with a pleasant odor. This reaction is an example of (1) saponification (2) esterification (3) polymerization (4) fermentation

13. Given the equation:

$$H-O-\underset{\underset{H}{|}}{\overset{\overset{H}{|}}{C}}-\underset{\underset{H}{|}}{\overset{\overset{H}{|}}{C}}-O-H + H-O-\underset{\underset{H}{|}}{\overset{\overset{H}{|}}{C}}-\underset{\underset{H}{|}}{\overset{\overset{H}{|}}{C}}-O-H \rightarrow H-O-\underset{\underset{H}{|}}{\overset{\overset{H}{|}}{C}}-\underset{\underset{H}{|}}{\overset{\overset{H}{|}}{C}}-O-\underset{\underset{H}{|}}{\overset{\overset{H}{|}}{C}}-\underset{\underset{H}{|}}{\overset{\overset{H}{|}}{C}}-O-H + H_2O$$

Which type of reaction is represented?
(1) condensation polymerization (2) addition polymerization
(4) esterification (4) saponification

14. Which is the formula of 2,2-dichloropropane?

15. The type of reaction represented by the equation
$C_2H_4 + H_2 \rightarrow C_2H_6$ is called (1) substitution (2) polymerization (3) addition (4) esterification

16. Which reaction is used to produce polyethylene $(C_2H_4)_n$ from ethylene? (1) addition polymerization (2) substitution (3) condensation polymerization (4) reduction

17. The process of opening double bonds and joining monomer molecules to form polyvinyl chloride is called (1) addition polymerization (2) condensation polymerization (3) dehydration polymerization (4) neutralization polymerization

18. Which hydrocarbon will undergo a substitution reaction with chlorine? (1) methane (2) ethyne (3) propene (4) butene

UNIT 10. APPLICATION OF CHEMICAL PRINCIPLES

CHEMICAL THEORY AND INDUSTRY

When scientists engage in what is termed "pure research", they are seeking knowledge for its own sake. The results of their quest for truth frequently are of great benefit to the quality of human life.

Industry is concerned with a maximum yield of products with maximum economic efficiency. The proper utilization of chemical principles materially affects quality, yield, and costs of production. Such utilization can only be accomplished through an understanding of these principles.

INDUSTRIAL APPLICATIONS

Equilibrium and Reaction Rates

Factors that affect rates of reaction, such as concentration, temperature, pressure, and catalysts are frequently the determinants of the processes by which many substances are manufactured. It is the application of some or all of these factors, as well as equilibrium conditions, that make many industrial processes practical.

The Haber Process is the name given to the industrial method of producing ammonia. This process involves the reaction:

$$N_2 + 3H_2 \leftrightarrows 2NH_3 + heat$$

The rate of formation of ammonia is increased by raising the temperature. However, since high temperatures favor endothermic reactions, the yield of ammonia would be reduced if the temperature were raised too much. In practice, therefore, a means of producing a compromise increase is utilized. Since the volume of the products is less than that of the reactants, increasing the pressure will produce a higher yield of product. The rate of ammonia production is further increased by adding iron and traces of aluminum oxide or silicon oxide along with potassium oxide, all of which serve as catalysts.

Contact process. The contact process is a multi-step procedure for producing sulfuric acid from sulfur or sulfide ores. The process starts with the burning of the initial material to produce sulfur dioxide.

$$S + O_2 \rightarrow SO_2 + heat$$

The sulfur dioxide is then oxidized to sulfur trioxide by the reaction

$$2SO_2 + O_2 \leftrightarrows 2SO_3 + heat$$

This reaction is catalyzed by the addition of platinum or vanadium pentoxide. The formation of SO_3 is slow at low temperatures. However, because high temperatures favor endothermic reactions, increasing the temperature would increase the rate of the reverse reaction and reduce the output of SO_3. In practice, temperatures between 450-575°C are maintained to produce reasonable yields. Increasing pressure also increases the yield.

The SO_3 gas is then absorbed by concentrated sulfuric acid.

$$H_2SO_4 + SO_3 \rightarrow H_2SO_4 \cdot SO_3, \text{ or } H_2S_2O_7$$

The product of this reaction is then diluted with water to produce sulfuric acid.

$$H_2S_2O_7 + H_2O \rightarrow 2H_2SO_4$$

Redox

Many industrial processes rely on redox reactions. Elements are extracted from their compounds and metals may be refined or protected through the application of the principles of oxidation and reduction.

Reduction of Metallic Compounds. The ores of metals tend to be extremely stable and are not very soluble in water. Since most of the metals in these ores are in their oxidized form, reduction of the ores is necessary. The method of reduction employed depends on the activity of the metal involved and the type of ore in which it occurs. The most active metals must be extracted from their compounds in the fused form by electrolytic processes. Groups 1 and 2 metals are obtained in this way, as described in unit VIII.

$$2NaCl(fused) \xrightarrow{\text{electricity}} 2\,Na + Cl_2$$

Some metals found in stable compounds are extracted from their ores by reacting them with metals that are stronger reducing agents. Aluminum is used in this way to extract chromium from its oxide.

$$2Al + Cr_2O_3 \rightarrow Al_2O_3 + 2Cr$$

Zinc and iron, which are considered moderately active metals, are extracted from their oxides by reduction by carbon in the form of coke, or by carbon monoxide. Sulfides and carbonates of some metals are first converted to their oxides and *then* subjected to this kind of reduction.

$$ZnO + C + heat \rightarrow Zn + CO$$

$$Fe_2O_3 + 3CO + heat \rightarrow 3CO_2 + 2\,Fe$$

Corrosion. Metals are often naturally reduced chemically to their oxides by the action of moisture and other chemical components of air. When the processes reduce or destroy the usefulness of the metal, the process is referred to as **corrosion**. Metals like aluminum and zinc form self-protecting coatings by virtue of the nature of their oxides. Once formed these oxides do not flake off as do those of iron (rust). Instead, they adhere tightly to the uncorroded metal beneath them and in this way prevent further corrosion.

Various methods are used to prevent the corrosion of metals like iron. Sometimes they are plated with self-protecting elements, such as aluminum or zinc (galvanizing), or with metals that resist corrosion, such as chromium or nickel. The alloying of iron with corrosion-resistant metals has produced stainless steel. Painting, coating with oils, or coating with glass (as in porcelain) are effective means of protection against corrosion.

Batteries. Batteries are electrochemical cells that provide electric energy from spontaneous redox reactions.

Lead-acid battery. In this widely used storage battery, the positive electrodes are made of lead oxide (PbO_2) and the negative electrodes are lead (Pb) in an electrolyte of sulfuric acid. The reaction is:

$$Pb + PbO_2 + 2H_2SO_4 \underset{\text{charge}}{\overset{\text{discharge}}{\rightleftharpoons}} 2PbSO_4 + 2H_2O$$

The concentration of the sulfuric acid is used to test these batteries. The concentration of this acid decreases as the battery discharges and increases when the battery is charged by reversing the reaction. The test most commonly utilized is an indirect one in which the specific gravity of the liquid is measured.

Nickel-cadmium battery. In these compact versions of the rechargeable storage battery, nickel hydroxide, $Ni(OH)_3$, is the positive electrode and cadmium (Cd) is the negative electrode. A solution of potassium hydroxide (KOH) provides the electrolyte. One of the advantages of this battery is that the concentration of the electrolyte does not change as the reaction proceeds.

$$2NiOOH + Cd + 2H_2O \underset{\text{charge}}{\overset{\text{discharge}}{\rightleftarrows}} 2Ni(OH)_2 + Cd(OH)_2$$

Note:—Several reactions are possible. This is the one believed most likely to occur.

Petroleum

The natural product known as petroleum is a complex mixture of hydrocarbons. It supplies the starting materials for many fuels, such as fuel oil, gasoline, propane, kerosene and butane. It also provides the raw materials for plastics, textiles, rubber, and detergents.

Natural gas (mostly methane), though not produced from petroleum, is a common fuel often found with petroleum deposits.

Processing Petroleum

Fractional distillation. Hydrocarbons of different boiling points are separated into fractions by distillation (evaporating, then condensing) at different temperatures. The more important fractions include gasoline, kerosene, fuel oil, lubricating oils and greases, paraffin wax, and asphalt.

Cracking. This process involves the heat-induced decomposition of large hydrocarbon molecules into smaller ones, usually with the help of catalysts. It is a very important process in the production of gasoline and fuel oil from crude oil. The oxides of aluminum and silicon are often the catalysts used.

QUESTIONS

1. When a battery is in use, stored chemical energy is first changed to
   (1) electrical energy (2) heat energy (3) light energy
   (4) mechanical energy

2. The reaction $CuO + CO \rightarrow CO_2 + Cu$ is an example of
   (1) reduction only (2) oxidation, only (3) both oxidation and reduction (4) neither oxidation nor reduction

3. Which group of metals is normally obtained by the electrolysis of their fused salts?
   (1) Group 17 (VIIA)   (2) Group 2 (IIA)   (3) Group 7 (VIIB)
   (4) Group 4 (IVB)

4. Which type of reaction is the Haber process,

$$N_2(g) + 3H_2(g) \rightarrow 2NH_3(g) + heat?$$

   (1) exothermic, with an increase in entropy
   (2) exothermic, with a decrease in entropy
   (3) endothermic, with an increase in entropy
   (4) endothermic, with a decrease in entropy

5. Kerosene is a mixture of compounds called
   (1) esters   (2) alcohols   (3) aldehydes   (4) hydrocarbons

6. Given the reaction at equilibrium:

$$C(s) + CO_2(g) + heat \leftrightharpoons 2CO(g)$$

   Which stress on the system would favor the production of $CO(g)$?
   (1) a decrease in the amount $C(s)$
   (2) a decrease in the amount of $CO_2(g)$
   (3) an increase in the pressure
   (4) an increase in the temperature

7. Given the equation for the discharge of a lead-acid battery:

$$Pb + PbO_2 + 2H_2SO_4 \rightarrow 2PbSO_4 + 2H_2O$$

   Which substance is oxidized?
   (1) Pb   (2) PbO_2 (3) H_2SO_4 (4) $PbSO_4$

8. Petroleum is a complex mixture of
   (1) alcohols   (2) ethers   (3) hydrocarbons   (4) organic acids

9. The process of cracking large hydrocarbon molecules produces
   (1) smaller molecules with higher boiling points
   (2) smaller molecules with lower boiling points
   (3) polymer molecules with higher boiling points
   (4) polymer molecules with lower boiling points

10. The equation below represents the reaction occurring in a nickel-cadmium battery.

$$2NiOOH + Cd + 2H_2O \rightarrow 2Ni(OH)_2 + Cd(OH)_2$$

Which reaction occurs at the cathode?
(1) reduction of Cd   (2) oxidation of Cd
(3) reduction of NiOOH   (4) oxidation of NiOOH

11. Which element is commercially obtained by the electrolysis of its fused salt? (1) silver (2) copper (3) helium (4) sodium

12. Natural gas is composed mostly of (1) butane (2) gasoline (3) methane (4) propane

13. The Haber process is used in the commercial preparation of (1) hydrochloric acid (2) sulfuric acid (3) ammonia (4) sulfur

14. The extent to which the metals zinc and aluminum corrode is limited because they (1) are semimetals (2) are amphoteric (3) form self-protective coatings by neutralization (4) form self-protective coatings by oxidation

15. Which is the negative electrode of a nickel oxide-cadmium battery? (1) Ni (2) Cd (3) $Ni(OH)_2$ (4) $Cd(OH)_2$

16. The electricity produced by a battery results from (1) an oxidation reaction, only (2) a reduction reaction, only (3) both an oxidation reaction and a reduction reaction (4) neither an oxidation reaction nor a reduction reaction

17. Petroleum is primarily a complex mixture of (1) hydrocarbons (2) esters (3) alcohols (4) organic acids

18. Which metals are produced commercially only by electrolysis of their fused salts? (1) Sr and Cr (2) Be and Fe (3) Li and Ni (4) Na and Ca

19. The corrosion of a metal is the result of a chemical reaction involving (1) oxidation, only (2) reduction, only (3) both oxidation and reduction (4) neither oxidation nor reduction

20. Given the balanced equation, which represents the Haber process:

$$N_2(g) + 3H_2(g) \leftrightarrows 2NH_3(g)$$

Which effect will an increase of pressure have on the system?
  (1) a decrease in the amount of NH_3 and a decrease in the amount N_2
  (2) a decrease in the amount of NH_3 and an increase in the amount N_2
  (3) an increase in the amount of NH_3 and a decrease in the amount N_2
  (4) an increase in the amount of NH_3 and an increase in the amount N_2

21. If the components of a liquid hydrocarbon are to be separated by fractional distillation, the components should have (1) large differences between their boiling points (2) very small differences between their boiling points (3) large differences between their freezing points (4) very small differences between their freezing points

22. One of the main products of the cracking of crude oil is (1) glycerol (2) gasoline (3) natural gas (4) nylon

23. A discharging lead-acid battery is best described as
  (1) chemical cells that use an electric current
  (2) chemical cells that produce an electric current
  (3) electrolytic cells that use an electric current
  (4) electrolytic cells that produce an electric current

24. Given the equation representing one step in the contact process:

$$X + O_2 \rightarrow \text{intermediate product}$$

Which compound is represented by X?
  (1) SO_2 (2) SO_3 (3) H_2SO_4 (4) $H_2S_2O_7$

25. Which metal will corrode, forming a protective layer that prevents further corrosion? (1) magnesium (2) sodium (3) aluminum (4) iron

UNIT 11. NUCLEAR CHEMISTRY

ARTIFICIAL RADIOACTIVITY

Natural radioactivity is the decay process occurring in elements that are radioactive as they exist in nature. This topic was addressed as a part of Unit 2, Atomic Structure. Radioactive elements can also be produced by bombarding the nuclei of stable atoms with high energy particles, such as protons, neutrons, and alpha particles. Such radioactivity is called **artificial radioactivity**. An example of artificial radioactivity is the nuclear reaction that takes place when the element beryllium is bombarded by protons.

$$^{9}_{4}\text{Be} + ^{1}_{1}\text{H} \rightarrow ^{6}_{3}\text{Li} + ^{4}_{2}\text{He}$$

Artificial Transmutation

When the nuclei of stable atoms are bombarded by accelerated particles, the nuclei become unstable and may cause the formation of isotopes or new elements. This process is called **artificial transmutation**. When this process occurs as the result of natural radioactivity, it is called **natural transmutation**. An example of artificial transmutation is the bombardment of the stable isotope Al-27 with an alpha particle:

$$^{27}_{13}\text{Al} + ^{4}_{2}\text{He} \rightarrow ^{30}_{15}\text{P} + ^{1}_{0}\text{n}$$

Accelerators. Particle accelerators of various types are used to give charged particles enough kinetic energy to overcome electrostatic forces and thus penetrate nuclei of target atoms. Acceleration is accomplished by means of the manipulation of electric and magnetic fields.

NUCLEAR ENERGY

Nuclear reactions involve energies that are millions of times greater than those found in ordinary chemical reactions. Energies of these magnitudes often are the result of the conversion of mass into energy.

Fission Reaction

Fission is the splitting of a heavy nucleus into two lighter nuclei. Only elements of high atomic number can be fissioned. The fission reaction is

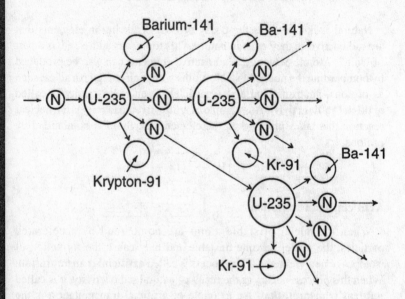

Figure 11-1. Nuclear Fission

brought about by the capture of neutrons. This results in the liberation of energy and the release of two or more neutrons per atom, in addition to the fission fragments. The energy liberated results from the conversion of mass into energy. The fact that each fission process emits more than one neutron is very important. The emitted neutrons induce other nuclei to split. In nuclear reactors, the *chain reaction* that results is controlled. In an atomic bomb, it is not controlled.

When heavy elements undergo fission, the new elements formed are more stable than the parent element because of the greater binding energy per nucleon.

Fission Reactors

A nuclear reactor is a device for controlling nuclear reactions so that the energy that is produced can be channeled to a useful form. Fission reactors produce energy from nuclear fission. The components of a fission reactor include fuel, moderators, control rods, coolants, and shielding.

Fuels. A nuclear fuel must be fissionable. Natural uranium consists of 99.3% U-238, which is *not* fissionable, and 0.7% U-235, which is fissionable. Thus, the U-235 is used as a nuclear fuel. An enriched form of uranium, which contains an additional 3-4% U-235 is also commonly used as fuel.

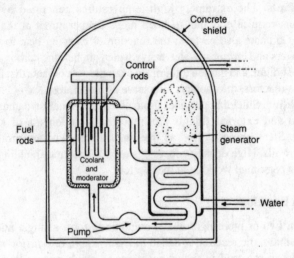

Figure 11-2. Nuclear reactor

Uranium-233 and plutonium-239 are also fissionable and are excellent fuels for reactors. U-233, produced from thorium-232, and plutonium-239, produced from uranium-238, are produced in special types of reactors called **breeder reactors**. Breeder reactors actually produce more fuel than they consume.

The world faces a uranium-235 shortage. Breeder reactors offer a means of using the much more abundant uranium-238 by converting it to fissionable plutonium-239. Thus, the breeder represents an almost unlimited supply of nuclear fuel. However, plutonium-239 is an extremely toxic substance. The danger presented by its toxicity tends to offset its favorable qualities as a plentiful fuel. For this reason, research is being focused on the production of uranium-233 from thorium-232. The uranium-233 is a much safer substance than is plutonium-239.

Moderators. Moderators are materials that have the ability to slow down fast-moving neutrons without absorbing them. This slowing down effect can be accomplished best by the head-on collision of the neutron with a particle of similar mass. Hydrogen and its isotope, deuterium, are such particles and have been found to be effective moderators. Commonly used moderators include water, heavy water, beryllium, and graphite.

Control Rods. The use of control rods provides a method of controlling the fission process by the absorption of neutrons. Thus, the number of neutrons available is controlled. The elements boron and cadmium absorb neutrons very well, and are therefore commonly used in control rods.

Coolants. The extremely high temperatures generated by fission reactions require coolants to keep these temperatures at reasonable levels. Coolants also perform the function of carrying heat to the heat exchangers and turbines. Water, heavy water, air, helium, carbon dioxide, molten sodium, and molten lithium are examples of coolants that are used. In some reactors, the coolants also serve as moderators.

Shielding. Shielding provides protection from radiation damage. Both internal and external shields are important components of a fission reactor. Steel lining is used as internal shielding, which protects the reactor walls. High density concrete is used as external shielding, which protects personnel working at the reactor.

Fusion Reaction

When two or more light nuclei combine to form a single nucleus of greater mass, the reaction is called *nuclear fusion,* or a *fusion reaction.* The energy released in fusion reactions is much greater than that in fission reactions. The mass of the new nucleus is *less* than the sum of the masses of the light nuclei. The difference in mass represents the mass that was converted to energy in the process. Some of this energy provides for the greater binding energy per nucleon and, therefore, the greater stability of the heavier and more stable nucleus formed.

Solar energy is believed to be the result of the fusion of ordinary hydrogen atoms into helium atoms. In order to achieve the energy necessary to initiate the fusion reaction of a hydrogen bomb, it is necessary to use a fission reaction.

Fuels. The fuels utilized for fusion reactions are the hydrogen isotopes deuterium, $_1^2H$, and tritium, $_1^3H$.

Deuterium may be obtained from heavy water, which can be extracted from ordinary water. Tritium is manufactured by the nuclear reaction:

$$_3^0Li + _0^1n \rightarrow _1^3H + _2^4He$$

High energy requirement. In order for nuclei to combine, they must have sufficient energy to overcome the forces of repulsion that exist between particles of like charge. The magnitude of the repulsive force increases with the size of the charge. Thus, only small nuclei, with very small charge, can be used in fusion reactions. Fusion with ordinary hydrogen, ^1_1H, is very slow. More rapid fusion reactions occur with the heavier hydrogen isotopes.

The idea of obtaining energy from fusion reactions is a very attractive one. Fusion reactors would use inexpensive, plentiful fuel and would produce practically no harmful radioactive wastes. However, there are some major problems that must be solved before practical fusion energy is a reality. These problems involve the extremely high activation energies necessary to trigger the fusion reaction. In order for fusion reactions to occur, temperatures on the order of 10^9 degrees C are required. The problems that must be solved are: How will these temperatures be achieved, and how will they be contained? The search for answers to these questions continues.

Radioactive Wastes

Fission products from nuclear reactors are highly radioactive and remain so for considerable periods of time. These highly dangerous substances cannot simply be discarded. Solid and liquid wastes, such as strontium-90 and cesium-137 are sealed in special containers that are stored underground or in isolated areas. Low-level radioactive wastes may be diluted and released into the environment. Gaseous radioactive wastes, such as radon-222, krypton-85, and nitrogen-16 are stored until they decay to safe levels, and then dispersed into the atmosphere.

USES OF RADIOISOTOPES

Major uses of radioactive isotopes can be classified into three basic groups:(1) uses based on chemical reactivity (2) uses based on radioactivity   (3) uses based on half-life.

Based on chemical reactivity. Fortunately, the chemical behavior of radioactive isotopes is the same as that of the stable isotopes of the same elements. As a result, radioactive isotopes can be used to trace the course of a reaction. The reaction continues in a normal fashion, but the path of the substituted radioactive material can be traced using radioactivity detection devices and methods. Many organic reaction mechanisms, including those in living systems, are studied using carbon-14 as the tracer.

Based on radioactivity. Isotopes with very short half-lives are administered to patients for diagnostic purposes. Tumors in various organs can be located and levels of activity of those organs can be monitored by administering and tracing radioactive substances known to concentrate in those organs. Technetium-99 is used to determine the location of brain tumors. Iodine-131 is used for the diagnosis of thyroid disturbances.

Cancer cells are more sensitive to radiation than are normal cells. Using carefully selected dosages, radioisotopes have been successfully used to treat cancer by destroying malignant cells. Dosages are sought that will attack only the malignant cells and not the normal cells. Radium and cobalt-60 are used in cancer therapy.

Radiation can be used to destroy bacteria, yeasts, and molds, as well as the eggs of insects. These capabilities provide the bases for the use of radioisotopes in food preservation.

Since the decrease in intensity of radiation is dependent on the amount of matter it passes through, radiation is used to measure thicknesses of industrial products. The wear-resistant properties of the moving parts of an engine may be measured by using radioactive parts and measuring the amount of radiation transferred to the lubricating materials used.

Based on half-life Radiochemical dating is a method of determining the age of fossils and rocks. The ratio of uranium-238 to lead-206 in a mineral can be used to determine the age of the mineral.

The half-life of a radioisotope is a constant factor. It is not affected by temperature, pressure, or any external factors. Thus, it can be assumed that a radioactive substance is decaying at the same rate today that it did at the time of its origin. The use of carbon-14 measurements in dating fossils relies on the fact that the activity of this isotope in living organisms is the same as it is in the atmosphere. It is assumed that the level of carbon-14 activity was the same when the fossil formed as it is today.

As long as an organism is living, the ratio of carbon-14 to carbon-12 is constant and remains constant. As soon as the organism dies, the carbon-14 lost through radioactive decay is not replaced. Thus, the ratio of the carbon-14 present in a fossil to that in the atmosphere can be used to determine the age of the fossil.

QUESTIONS

1. A positively charged particle has great difficulty penetrating a target nucleus because the charged nucleus has
   (1) a positive charge, which repels the particle
   (2) a negative charge, which attracts the particle
   (3) the protection of surrounding electrons
   (4) a very high binding energy

2. Which radioactive waste can be stored for decay and then safely released directly into the environment? (1) N-16 (2) Sr-90 (3) Cs-137 (4) Pu-239

3. Which equation represents a nuclear reaction that is an example of artificial transmutation?

   (1) $^{43}_{21}Sc \rightarrow ^{43}_{20}Ca + ^{0}_{+1}e$ (2) $^{14}_{7}N + ^{4}_{2}He \rightarrow ^{17}_{8}O + ^{1}_{1}H$

   (3) $^{10}_{4}Be \rightarrow ^{10}_{5}B + ^{0}_{-1}e$ (4) $^{14}_{6}C \rightarrow ^{14}_{7}N + ^{0}_{-1}e$

4. In the construction of some nuclear reactors, a radiation barrier of concrete is used as the (1) external shield to protect reactor walls (3) internal shield to protect reactor walls (3) external shield to protect personnel (4) internal shield to protect personnel

5. The operation of a commercial nuclear reactor in New York State requires an isotope that will undergo (1) fission and controlled chain reaction (2) fission and an uncontrolled chain reaction (3) fusion and a controlled chain reaction (4) fusion and an uncontrolled chain reaction

6. Which particle can *not* be accelerated by the electric or the magnetic field in a particle accelerator? (1) electron (2) neutron (3) helium nucleus (4) hydrogen nucleus

7. A radioactive-dating procedure to determine the age of a mineral compares the mineral's remaining amounts of isotope ^{238}U and isotope (1) ^{206}Pb (2) ^{206}Bi (3) ^{214}Pb (4) ^{214}Bi

8. The fission process in a reactor can be regulated by adjusting the number of neutrons available. This is done by the use of (1) moderators (2) control rods (3) coolants (4) shielding

9. Particle accelerators can be used to increase the kinetic energy of (1) deuterium (2) neutrons (3) protons (4) tritium

10. Given the reaction $^{7}_{3}Li + X \rightarrow ^{8}_{4}Be$

    Which species is represented by X?

    (1) $^{1}_{1}H$ (2) $^{2}_{1}H$ (3) $^{3}_{2}He$ (4) $^{4}_{2}He$

11. Radioisotopes used for medical diagnosis must have (1) long half-lives and be quickly eliminated by the body (2) long half-lives and be slowly eliminated by the body (3) short half-lives and be quickly eliminated by the body (4) short half-lives and be slowly eliminated by the body

12. Compared to an ordinary chemical reaction, a fission reaction will
    (1) release smaller amounts of energy
    (2) release larger amounts of energy
    (3) absorb smaller amounts of energy
    (4) absorb larger amounts of energy

13. The primary purpose of the moderator used in a nuclear reactor is to (1) absorb protons (2) absorb neutrons (3) slow protons (4) slow neutrons

14. Which fissionable elements are produced in breeder reactors?
    (1) lithium-6 and hydrogen-3
    (2) carbon-14 and oxygen-17
    (3) uranium-233 and plutonium-239
    (4) cesium-137 and radon-222

15. Given the equation: $^{14}_{7}N + ^{4}_{2}He \rightarrow X + ^{17}_{8}O$

    When the equation is correctly balanced, the particle represented by the X will be

    (1) $^{0}_{-1}e$ (2) $^{1}_{0}n$ (3) $^{1}_{1}H$ (4) $^{2}_{1}H$

16. Which substance may serve as both a moderator and coolant in some nuclear reactors? (1) carbon dioxide (2) boron (3) graphite (4) heavy water

17. Iodine-131 is used for diagnosing thyroid disorders because it is absorbed by the thyroid gland and (1) has a very short half-life (2) has a very long half-life (3) emits alpha radiation (4) emits gamma radiation

18. Which element is sometimes used as a moderator in a nuclear reactor? (1) C   (2) Cu   (3) Al   (4) Zn

19. Given the nuclear reaction: $_6^{12}C + _1^2H \rightarrow X + _0^1n$

When the equation is correctly balanced, the nucleus represented by the X is (1) $_6^{14}N$ (2) $_7^{13}N$ (3) $_7^{13}C$ (4) $_6^{13}C$

20. In a fission reaction uranium-235 is used as a
(1) coolant (2) moderator (3) fuel (4) control rod

21. The waste products from nuclear reactors can be in the form of
(1) solids, only (2) solids and liquids, only (3) solids and gases, only (4) solids, liquids, and gases

22. Which radioactive isotope is often used as a tracer to study organic reaction mechanisms?
(1) Carbon-12 (2) carbon-14 (3) uranium-235 (4) uranium-238

23. In the reaction $_{13}^{27}Al + _2^4He \rightarrow X + _0^1n$, the isotope represent by X is

(1) $_{12}^{29}Mg$ (2) $_{13}^{28}Al$ (3) $_{14}^{27}Si$ (4) $_{15}^{30}P$

UNIT 12. LABORATORY ACTIVITIES

SIGNIFICANT FIGURES

Many operations in the chemistry laboratory involve measurements. Finding mass and volume, for example, both require measurement operations. Recording the results of these measurements should be based on certain rules in order to show the precision of the measurement.

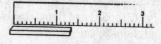

Figure 12-1.

All measurements involve uncertainty which is ordinarily recorded in the last digit. For example, the length of the object shown being measured is between 1.3 and 1.4 units. Its length may be recorded as 1.35 units. The last digit, 5, is estimated, but it is considered to be significant and should be recorded. In interpreting a recorded mass of 4.7 grams, the assumption would be that the object was massed to the nearest tenth of a gram and that its exact mass is between 4.65 grams and 4.75 grams. If the recorded mass of the object were 4.733 g, there are *four* significant figures. It must be assumed that the object was massed to the nearest thousandth of a gram. Significant digits consist of those that we know with certainty *plus* one more doubtful, or estimated, digit.

Zeros may or may not be significant, according to a set of conventional rules. The number 6.00 has three significant figures, the two zeros being significant. This number implies that the measurement was made to the nearest hundredth. The zeros in 4.023, 6.105, 8.210, 20.10, and 3.002 are all considered significant. Each of these numbers contains four significant figures. One should always record all quantities provided by the measurement device used. An object massed as exactly two grams on a balance that is accurate to the nearest hundredth of a gram should be recorded as 2.00g, *not* 2g.

When a zero is used to locate a decimal point, as in 0.023, the zero is *not* a significant figure. The numbers 0.036, 0.00087, and 0.0020 have only *two* significant figures each. These numbers could be written in scientific notation, 3.6×10^{-2}, 8.7×10^{-4}, and 2.0×10^{-3}. This is one of the advantages of the use of scientific notation.

Calculations and Significant Figures

In any calculation in which experimental results are used, the final result should contain only as many significant figures as are justified by the instruments employed in finding those results. The measurement of *least* precision determines the number of significant figures in the final answer.

Addition and subtraction. In addition and subtraction, retain only as many *decimal places* in the result as there are in that item which uses the *least* number of decimal places.

Example: 42.2   (limiting measurement)
                                       3.024
                                  +   7.23
                                                    Answer: 52.5

Multiplication and division. In multiplication and division problems, the answer should contain only as many significant figures as are contained in the item with the *least* number of significant figures.

Example: $43.42 \times 0.029 \times 44.6 = 56.$

Since the number 0.029 has only two significant figures, the answer must be 56., which contains two significant figures.

LABORATORY SKILLS

Some equipment commonly used in the chemistry laboratory are illustrated in Figure 12-3. You should be able to identify each piece and describe how it is used.

Using the Balance.

The main types of balance found in the high school laboratories are the electronic balance, the "dial" type balance, and the beam balance. The beam balance remains the "mainstay" of any laboratory. The beam balance may look different than the one shown here, but all balances have several features in common.

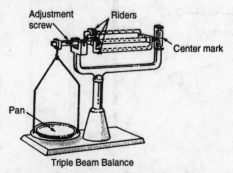

Figure 12-2. Triple-beam balance

Figure 12-3.

The following steps should be followed in using a beam balance.

1. Slide all riders back to the zero point. Check to see that the pointer swings freely along the scale. The beam should swing an equal distance above and below the zero point. If necessary, use the adjustment screw to obtain an equal swing of the beams. You should "zero" the balance every time you use it.

2. Never put a hot object directly on the balance pan. Air currents developing around the hot object may cause massing errors.

3. Never pour chemicals directly on the balance pan. Dry chemicals should be placed on paper or in a glass container. Liquid chemicals should be massed in glass containers.

4. Place the object to be massed on the pan and move the riders along the beams, beginning with the largest mass first. If the beams are notched, make sure all riders are in a notch before you take a reading.

5. The mass of the object will be the sum of the masses indicated on the beams. Subtract the mass of the paper or container from the total mass reading.

Using the Gas Burner

There are several different types of laboratory gas burners. While these burners differ somewhat in appearance, most have similar construction. Figure 12-4 shows a typical gas burner.

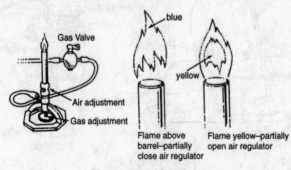

Figure 12-4. Gas burner

The gas outlet jet is in the "off" position when the handle is at a right angle to the outlet. To light the burner, hold a lighted match or striker near the top of the burner barrel. Turn the gas outlet jet handle to the "on" position. After the gas has ignited, turn the air adjustment vent until a light blue cone appears in the center of the flame. If the flame rises from the burner barrel or appears to blow out, adjust the gas-air mixture. Some burners have a knob at the bottom of the burner for this purpose. If no knob is present, adjust the flow of gas from the gas outlet jet.

Decanting and Filtering

It is often necessary to separate a solid precipitate from a liquid solution. The most common process of separation is a two-step one—decanting and then filtering. Decanting is the careful pouring off of most of the liquid, making sure that all of the solid remains behind. To avoid splashing during this procedure, pour the liquid down a stirring rod, as shown in part A of Figure 12-5.

The first step in filtration is to prepare the filter paper. Fold the paper as shown in part B of Figure 12-5. Open up the filter, place it in the funnel, and wet it with distilled water from a wash bottle. Press the paper into position in the funnel making sure the filter lies flat against the sides of the funnel.

The beaker that is to receive the filtrate should be positioned so that the longer side of the funnel stem lies along the side of the beaker, as shown in part C of Figure 12-5. After all the liquid has been poured from the beaker, use a wash bottle to remove any remaining solid particles. Also wash the solid collected in the filter before removing the filter from the funnel.

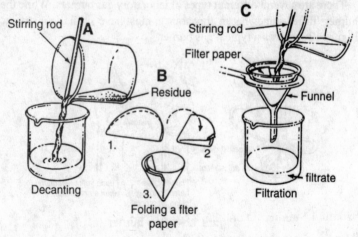

Figure 12-5. Decanting and filtering

Working with Glass Tubing

When setting up various apparatus in the chemistry laboratory, it is often necessary to cut, polish, and bend pieces of glass tubing, and to insert the tubing into rubber stoppers.

Cutting glass tubing. Place the tubing to be cut on a flat surface. Using a triangular file, scratch the tubing at the point where you are going to break it. To make this scratch, hold the tubing firmly with one hand and make one firm stroke away from you with the file.

Pick up the tubing and hold it with the scratch facing away from you. Place your thumbs on either side of the scratch and push firmly with your thumbs, as shown in Figure 12-6.

scratch

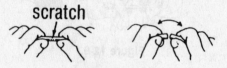

Figure 12-6.

Fire polishing. The broken ends of the glass tubing are quite sharp and should be polished. Place the cut end of the glass in a burner flame and rotate it back and forth until the flame becomes bright yellow. At this point, the rough edges should be smooth. Remove the glass from the flame before the opening starts to close up. Place the glass on an insulated pad to cool. *Caution:* Hot glass can cause painful burns.

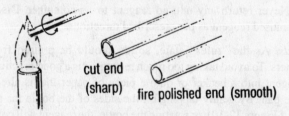

cut end (sharp)   fire polished end (smooth)

Figure 12-7. Fire polishing glass tubing

Bending glass tubing. Use a flame spreader on your gas burner to bend glass tubing. Hold the glass tubing in the flame as shown in Figure 12-8.

Figure 12-8.

Rotate the glass back and forth until the flame becomes yellow and the glass softens. Remove the softened glass from the flame and lift both ends of the glass in a smooth, quick motion, producing a smooth curve.

Inserting glass tubing into a stopper. Lubricate the hole of the stopper and the end of the tubing with water or glycerine. Hold the tubing with a piece of toweling. Insert the tubing into the stopper. When inserted, the tubing should protrude from the stopper, as shown in part B of Figure 12-9.

Figure 12-9.

Transferring Chemical Reagents

To maintain the purity of a chemical reagent, the following rules should always be followed:

- Whenever possible, solid reagents should be poured from their containers.
- If necessary to scoop reagent from its container, use a clean spoon or spatula that is reserved exclusively for use with that reagent.
- Never return any unused reagent to its container. Discard any unused reagent as per prescribed directions.

Solids. As the "rules" state, solids should be poured from their containers. To avoid taking too much reagent, it is a good practice to pour the reagent into a beaker or onto a piece of paper that is already on a balance pan. By gently tapping on the sides of the bottle (as shown in part A of Figure 12-10), or rotating the bottle, the reagent will come out in a gentle flow.

If the reagent is poured onto a piece of paper, the paper should be creased and used to transfer the reagent as shown in Figure 12-10B.

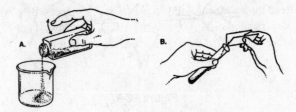

Figure 12-10.

Liquids. When transferring liquids from a reagent bottle, remove the stopper as shown in Figure 12-11. Keep the stopper between your fingers throughout the transfer. *Do not set it down.*

Figure 12-11.

When pouring a liquid into a wide-mouth container, pour it down a stirring rod, as shown in part A to Figure 12-12, to avoid splashing. When pouring into a narrow-mouth container, such as a test tube, hold the test tube with a test tube holder and pour as shown in part B of Figure 12-12.

Figure 12-12.

Reading Liquid Measurements

Graduated cylinders, pipets and burets are the pieces of equipment most often used for measuring volumes of liquids. They may be made of glass or some type of plastic material. Surfaces of liquids in cylindrical containers are always curved. The curved surface is called a *meniscus.* Most liquids will produce a meniscus that is curved downward in the middle. When reading the volume of such liquids two important rules must be followed:

- Read the volume from the scale on the container where it matches the central part of the meniscus (usually the lowest part).
- Make certain that the reading is made at eye level.

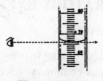

Figure 12-13.

All measurements should be recorded using the rules of significant figures. This means recording the units written and/or marked on the measuring device plus an additional estimated unit. Volume-measuring instruments may be calibrated "to deliver" the volumes marked on their scales, or "to contain" the volumes of their scales. "To contain" scales have zero marked at the bottom. "To deliver" scales have zero marked at the top. A convention followed by most manufacturers is to use red markings or a vertical red band on "to contain" cylinders. Sometimes, the letters "TC" are marked on them.

Graduated cylinders are most often marked with the zero on the bottom of the. scale. Burets will often have the zero at the top. See Figure 12-14 for proper reading of the two types of scales.

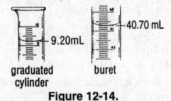

graduated   buret
cylinder
Figure 12-14.

Using Thermometers.

Most laboratory thermometers are mercury or alcohol-filled and measure temperatures from -20°C to 110°C. The scales are usually graduated in units of 1 C°. Readings, therefore, should be taken to the nearest degree and recorded to the nearest tenth of a degree, estimating the tenth place.

Some thermometers are designed to be read with the thermometer totally immersed in the substance whose temperature is being measured. These have the words "total immersion" marked on them. Other thermometers are marked to indicate the extent to which the thermometer should be immersed in order to get the most accurate readings. The rest of the thermometer remains at room temperature. When measuring liquids while they are being heated or cooled, it is important that the liquid be kept in motion and the thermometer is not touching the bottom or walls of the container.

LABORATORY EXERCISES

Heating and Cooling Curves

Liquid-solid phase change, finding melting points of pure substances, and constructing heating and cooling curves can all be accomplished by means of an exercise using paradichlorobenzene or naphthalene. Place either of these two substances in a test tube and heat the contents to a temperature not to exceed 100°C. Remove from heat source. Then, take temperature readings at one-minute intervals until the temperature remains constant for several successive readings. Take a few more readings as the temperature begins to fall. This provides data for a cooling curve. The process may be reversed to produce a heating curve. The "plateau," or flat part of the curve should indicate the melting or freezing point of the substance.

Simple Calorimetry

Finding the heat of combustion of candle wax provides a good example of how water can be used as a calorimetric liquid. The mass of a candle and the mass and temperature of a sample of water are carefully measured. The candle is then used to heat the water for a carefully measured amount of time. At the end of the heating period, the mass of the candle and the temperature of the water are measured again. The heat (calories) absorbed by the water is found by using the equation

calories = mass × specific heat × temperature change

recalling that the specific heat of water is 1 cal/g-C°. This number of calories, divided by the mass *lost* by the candle, indicates the heat of combustion of the candle in cal/g.

SAMPLE PROBLEM

A candle with a mass of 35.25 grams is used to heat 235 grams of water at a temperature of 22.8°C. After heating for several minutes, the candle is extinguished. The final mass of the candle is 34.10 grams. The final temperature of the water is 64.3°C. Find the heat of combustion of the candle.

Solution:
1. Determine the mass of the candle that burned.

    original mass – final mass = mass lost

    35.25g – 34.10 g = 1.15 grams lost
2. Determine the temperature change of the water.

    Δt = 64.3°C - 22.8°C

    Δt = 41.5°C
3. Determine the number of calories of heat produced.

    calories = mass × specific heat × Δt

    = 235 g × 1 cal/g°C × 41.5°C

    = 9750 calories
4. Calculate the heat of combustion (cal/g).

$$\frac{9750}{1.15g} = \textbf{8480 cal/g}$$

Endothermic and Exothermic Processes

The heat of dissolution of substances can be determined by using a simple calorimeter. A measured mass of water is placed in a styrofoam cup. A measured mass of ammonium chloride crystals is prepared and the temperature of the water in the cup is measured. The ammonium chloride is added to the water and the temperature of the mixture is read constantly until total solution is accomplished. The number of calories consumed (endothermic) is determined, using the mass of the water, its specific heat, and its change in temperature. The heat of dissolution is determined in cal/g by dividing the number of calories consumed by the mass of the ammonium chloride. This value can be converted to cal/mole. Moles of ammonium chloride are calculated by multiplying the mass in grams times the conversion factor mole ammonium chloride/gram molecular mass of ammonium chloride.

SAMPLE PROBLEM

When 5.4 grams of ammonium chloride are dissolved in 100. g of water, the temperature of the water decreases from 30.0° C to 26.0° C. What is the heat of dissolution?

Solution:

1. Find the number of calories represented by this temperature change.

$$\text{calories} = \text{mass of water} \times \text{specific heat} \times \Delta t$$
$$= 100. \text{ g} \times 1 \text{ cal/g°C} \times 4.0°C$$
$$= 4.0 \times 10^2 \text{ calories}$$

2. Heat of reaction (dissolution) is expressed in cal/mole.
   Convert grams NH_4Cl to moles NH_4Cl.

$$5.4g \times \frac{1 \text{ mole}}{53.5g} = 0.10 \text{ mole } NH_4Cl$$

3. Find the heat of dissolution in kcal/mole.

$$\frac{4.0 \times 10^2}{0.10 \text{ mole}} = 4.0 \times \frac{10^3 \text{ cal}}{\text{mole}} = 4.0 \text{ kcal/mole}$$

Solubility Curve

Solubility curves are most often produced by plotting the concentration of a saturated solution (in g solute/100g solvent) against the temperature at which saturation occurs. As a laboratory exercise, potassium nitrate is a good choice of solute. A predetermined mass of solute is placed in a test tube and a carefully measured volume of water added. The mixture is heated gently until solution is complete. The solution is then allowed to cool with a thermometer in it, so that the temperature at which crystallization begins can be read and recorded. The procedure is then repeated, using increasing masses of solute in the same volume of water. In processing the data, the volume of water is converted to grams. The values plotted on the y-axis are g sol/100g water and the temperature is plotted along the x-axis.

Differences Between Organic and Inorganic Substances

Using sodium chloride as the model inorganic substance and sucrose as the model organic substance, the results for each in the following may be compared:

- Melting point determination (Apply heat very slowly.)
- Electrical Conductivity
- Stability can be compared by heating both rapidly enough to "caramelize" the sugar. It will be seen to turn brown, while the sodium chloride will not.
- Solubility in various solvents. To observe this difference it would be better to use different substances and compare the solubilities of several organic and several inorganic substances in the same polar and nonpolar solvents.

Flame Tests

Metallic ions can be identified by the use of flame tests. Flame tests are made using a platinum or nichrome wire loop sealed in the end of a piece of glass tubing (or other suitable handle). The loop is cleaned by dipping it in hydrochloric acid and then heating it in the flame of a bunsen burner. This should be repeated until the wire no longer imparts a color to the flame. The wire loop is then dipped into a solution and placed back into the flame. One at a time, solutions known to contain the ions of lithium, sodium, potassium, calcium, strontium and copper should be subjected to flame tests and the colors recorded. The loop should be cleaned after each test. Next, the results of tests with unknown solutions can be compared to the standard colors to identify the unknowns. Colors usually found are: lithium-deep red; sodium-yellow; potassium-violet; calcium-yellowish red, strontium-bright red; and copper-green.

Water of Hydration

Many ionic solids contain water within their crystalline structure in a fixed molar ratio. The water is included in the empirical formulas: $BaCl_2 \cdot 2H_2O$ and $CuSO_4 \cdot 5H_2O$ are examples. In this exercise, the percentage, by mass, of water in the hydrated substance can be determined. The mass of a sample is accurately measured and then carefully heated to drive off the water. The anhydrous (without water) salt is cooled and massed. The difference in mass of the sample before and after heating represents the mass of the water of hydration. Percentage is calculated as follows:

$$\% \text{ water} = \frac{\text{mass of water of hydration}}{\text{mass of hydrated salt}} \times 100\%$$

From the same set of data, the empirical ratio of water to ionic solid can be calculated by converting the mass values to mole values. This ratio provides the coefficient of H_2O in the empirical formula of the hydrated salt.

Molar Volume of a Gas

The hydrogen gas generated when approximately 0.04 g magnesium is reacted with excess dilute hydrochloric acid can be collected in a 50 mL buret or a eudiometer tube (25- or 50-mL). The volume of the hydrogen gas should be corrected to STP. The pressure inside the tube must be adjusted to atmospheric pressure. This is done by reading the volume of the gas while the tube is inverted in a column of water. The surface inside the gas tube must be level with the surface of the water outside the tube In addition, the temperature of the water must be measured. Then, the vapor pressure of the water at that temperature must be subtracted from atmospheric pressure to find the pressure of the hydrogen gas alone. By converting the mass of the magnesium to moles, and using the mole ratio of magnesium to hydrogen in the balanced equation for the reaction, the number of moles of hydrogen gas can be calculated. The volume of the hydrogen gas collected (corrected to STP) divided by the number of moles of hydrogen will equal the molar volume (volume of 1 mole at STP) of hydrogen.

SAMPLE PROBLEM

A mass of 0.0520 g of magnesium is reacted with HCl to produce 51.8 mL of hydrogen gas at a pressure of 755 torr and a temperature of 17° C. Determine the molar volume of hydrogen gas from this data.

Solution:

1. Convert grams of Mg to moles of Mg.

$$\frac{0.0520 \text{ g}}{24.3 \text{ g/mole}} = .00214 \text{ moles Mg}$$

2. Correct for vapor pressure of water.

$$P_{H_2} = P_{total} - P_{H_2O}$$

According to Reference Table O, the vapor pressure of water at 17°C is 15 torr

$$P_{H_2} = 755 \text{ torr} - 15 \text{ torr}$$
$$= 740. \text{ torr}$$

3. Calculate the volume of H_2 gas at STP. (combined gas law)

$$V_{STP} = \frac{51.8 \text{ mL} \times 740. \text{ torr} \times 273 \text{ K}}{760 \text{ torr} \times 290. \text{ K}}$$
$$= 47.3 \text{ mL}$$

4. Calculate the volume occupied by one mole (molar volume).

$$\frac{47.3 \text{ mL}}{0.00214 \text{ mole}} = \frac{x \text{ mL}}{1 \text{ mole}}$$
$$\mathbf{22\ 100 \text{ mL} = x}$$

Percent Error

All experiments involve some error. Consider, for example, the sample problem just described. Based on the "experimental data," the molar volume of hydrogen gas at STP was determined to be 22 100 milliliters. The accepted value for the molar volume of any gas at STP is 22 400 milliliters. The error in this instance is quite small.

The percent error of any given or measured value is a comparison of that value with the number that is accepted as being the true value. The following expression can be used to find percent error:

$$\% \text{ error} = \frac{\text{difference between accepted and experimental values}}{\text{accepted value}} \times 100\%$$

The percent error for the sample problem data is

$$\frac{22\ 400\ \text{mL} - 22\ 100\ \text{mL}}{22\ 400\ \text{mL}} = \times 100\% = 1\%$$

Titration

Titration is a method of determining the concentration of a solution of unknown concentration by reacting it with a standard solution of known concentration. Titration is commonly employed in neutralization reactions between acids and bases. A carefully measured volume of acid (or base) of unknown concentration is placed in an appropriate container. A few drops of an indicator is added to this solution. The indicator is one that will change color at the point at which neutralization is reached. This is called the *endpoint* of the titration. A standard base (or acid) of known concentration is delivered into the solution of unknown concentration from a buret. The standard is delivered slowly and the container holding the unknown is constantly agitated. The standard solution is added until the color change is complete. The volume of standard solution used to achieve neutralization is carefully noted. The unknown concentration is calculated as follows:

volume of unknown × X = volume of standard × concentration of standard where X is the unknown concentration. This equation can be modified if the mole ratio in the reaction is not 1:1.

Good results can be achieved in titrating dilute solutions of sodium hydroxide or potassium hydroxide against dilute solutions of hydrochloric or sulfuric acid, using methyl orange or phenolphthalein as the indicator.

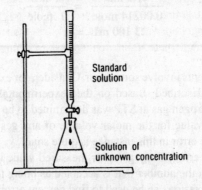

Standard
solution

Solution of
unknown concentration

Figure 12-15.

Laboratory Reports

Performing laboratory exercises is an important part of the study of chemistry. The manner in which reports of these exercises are written is equally important. Laboratory reports should contain all the important information written in a clear and concise format.

While specific formats may be used, they should all provide students with the opportunity to demonstrate their ability to:

1. Organize information in a logical manner.
2. Record data properly in the form of tables and graphs whenever possible.
3. Make a list of observations.
4. Draw conclusions based on observations.
5. Calculate percentage error whenever quantitative findings are involved with respect to known or accepted values.

QUESTIONS

1. The solid object shown below has a mass of 162.2 grams.

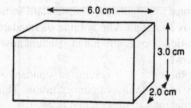

What is the density of the object to the correct number of significant figures? (1) 0.22 g/cm³ (2) 0.2219 g/cm³ (3) 4.5 g/cm³ (4) 4.505 g/cm³

2. When 5.0 grams of solute is added to 25.0 grams of water at 35.0°C, the temperature rises to 35.4°C. What is the temperature change per gram of solute, expressed to the correct number of significant figures? (1) 0.08 C°/g (2) 0.080 C°/g (3) 8.0 C°/g (4) 8.00 C°/g

3. A student wishes to mix 25 milliliters of 12 M hydrochloric acid and 75 milliliters of distilled water. Which procedure is safest? (1) adding all the acid to the water and then slowly stirring the mixture (2) adding all the water to the acid and then slowly stirring the mixture (3) stirring constantly while slowly adding the acid to the water (4) stirring constantly while slowly adding the water to the acid

4. A student obtained the following data while cooling a substance.
    The substance was originally in the liquid phase at a temperature
    below its boiling point.

Time (minutes)	0.5	1.0	1.5	2.0	2.5	3.0	3.5	4.0	4.5	5.0	5.5	6.0
Temperature (°C)	70.	63	57	54	53	53	53	53	53	52	51	48

What is the freezing point of the substance?   (1) 70.°C   (2) 59°C
(3) 53°C   (4) 48°C

5. A student observing the behavior of paradichlorobenzene first heats
    10 grams of the substance in a hot water bath until it is completely
    liquefied. The following data are recorded as paradichloroben-
    zene cools.

DATA TABLE

Time (minutes)	Temperature (°C)
0	65
1	58
2	52
3	53
4	53
5	53
6	53
7	53
8	51
9	47
10	42

What is the freezing point of paradichlorobenzene?
(1) 42°C   (2) 53°C   (3) 58°C   (4) 65°C

6. A student is filtering a mixture of sand and salt water into a beaker.
    What will be found in the beaker after the filtration is completed?
    (1) sand, only   (2) salt, only   (3) sand and salt   (4) salt and water

7. In a laboratory exercise to determine the density of a substance, a
    student found the mass of the substance to be 6.00 grams and the
    volume to be 2.0 milliliters. Expressed to the correct number of
    significant figures, the density of the substance is   (1) 3.000 g/mL
    (2) 3.00 g/mL   (3) 3.0 g/mL   (4) 3 g/mL

8. A 1.20-gram sample of a hydrated salt is heated to a constant mass of 0.80 gram. What was the percent by mass of water contained in the original sample? (1) 20.   (2) 33   (3) 50.   (4) 67

9. The graph below represents the decay of a radioactive isotope.

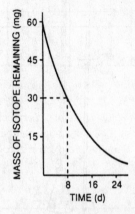

Based on Reference Table H, which radioisotope is best represented by the graph? (1) ^{32}P   (2) ^{131}I   (3) ^{198}Au   (4) ^{222}Rn

10. In an experiment, a student found the percent by mass of water in a sample of $BaCl_2 \cdot 2H_2O$ to be 17.4%. If the accepted value is 14.8%, the percent error in the student's result is equal to

(1) $\dfrac{2.6}{17.4} \times 100$ (2) $\dfrac{2.6}{14.8} \times 100$

(3) $\dfrac{14.8}{17.4} \times 100$ (4) $\dfrac{17.4}{14.8} \times 100$

11. What is the sum of 0.0421 g + 5.263 g + 2.13 g to the correct number of significant digits? (1) 7 g   (2) 7.4 g   (3) 7.44 g (4) 7.435 g

12. Which diagram represents a graduated cylinder?

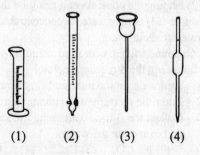

(1)   (2)   (3)   (4)

13. Which laboratory procedure uses the equipment shown in the diagram below?

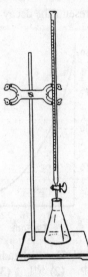

(1) evaporation   (2) condensation   (3) distillation   (4) titration

14. Which statement explains why mass is lost when a student heats a sample of $BaCl_2 \cdot 2H_2O$ crystals?   (1) Chlorine is given off as a gas.   (2) Water is given off as a gas.   (3) The crystals sublime.   (4) The crystals fuse (melt).

15. The mass of a solid is 3.60 grams and its volume is 1.8 cubic centimeters. What is the density of the solid, expressed to the correct number of significant figures?   (1) 2 g/cm^3 (2) 2.0 g/cm^3 (3) 0.5 g/cm^3 (4) 0.50 g/cm^3

16. Which activity is considered a proper laboratory technique? (1) heating the contents of an open test tube held vertically over a flame   (2) heating the contents of a test tube that has been closed with a stopper   (3) adding water to concentrated acids   (4) adding concentrated acids to water

17. Which compound mixed with sand could be separated from the sand by following the three steps below?

Step 1 – Add water to the mixture of the compound and sand.

Step 2 – Filter the mixture of the compound and sand.

Step 3 – Collect the filtrate containing the soluble component and evaporate the water.

(1) $BaCO_3$ (2) Na_2CO_3 (3) $HgCl$ (4) $AgCl$

18. The graph below represents four solubility curves. Which curve best represents the solubility of a gas in water?

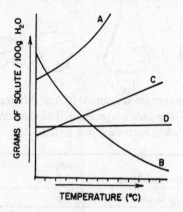

(1) A   (2) B   (3) C   (4) D

19. The graph below represents the relationship between the vapor pressure and temperature of four liquids.

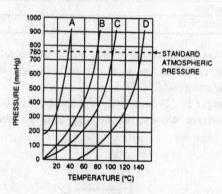

Which liquid has a normal boiling point of 79°C?
(1) A   (2) B   (3) C   (4) D

20. Which represents an Erlenmeyer flask?

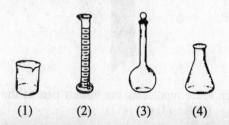

(1)   (2)   (3)   (4)

21. Which diagram represents a crucible?

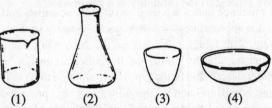

(1)	(2)	(3)	(4)

22. A student investigated samples of four different substances in the solid state. The table is a record of the behaviors observed (marked with an *X*) when each solid was tested.

Behavior Tested	Sample I	Sample II	Sample III	Sample IV
High Melting Point	X		X	
Low Melting Point		X		X
Soluble in Water	X			X
Insoluble in Water		X	X	
Decomposed Under High Heat		X		
Stable Under High Heat	X		X	X
Electrolyte	X			X
Nonelectrolyte		X	X	

Based on the tabulated results, which of the solids investigated had the characteristics most closely associated with those of an organic compound?

(1) Sample I   (2) Sample II   (3) Sample III   (4) Sample IV

23. The diagram below shows the upper part of a laboratory burner.

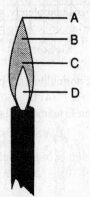

Which letter represents the hottest part of the burner flame?

(1) A   (2) B   (3) C   (4) D

REFERENCE TABLES FOR CHEMISTRY

Since reference tables are used so often by scientists and students of science, it is appropriate that high school chemistry students be familiar with them. The tables provided by the New York State Education Department for students taking the Regents Examination are presented here. A brief description of the type of information to be found and some test items based on each of the tables are also provided. However, students are urged to peruse the entire booklet of reference tables thoroughly while taking the examination. There is much additional information to be found in many of the tables. For example, tables may contain definitions, sample equations, the names and the formulas of ions and compounds, etc.

Some Regents test items contain phrases such as: "according to Reference Table L . . . ," or "based on Reference Table K." Many other items require information contained in the tables. Students should be aware of this. Competency in information retrieval is one of the important objectives of the chemistry program.

TABLE A

Physical Constants and Conversion Factors

The constants listed in Table A are very useful in processing laboratory data and in many calculations. Competency in using most of Table A is not tested on most Regents Examinations.

Freezing point depression and boiling point elevation are effects on solutes that depend on the number of particles in solutions. Therefore, ionic substances produce greater effects than do covalent substances. For example, since NaCl ionizes completely, a one molal solution produces two moles of particles per liter (one mole Na^+, one mole Cl^-). As a result, the freezing point would be depressed $3.72C°$ and the boiling point elevated $1.024C°$. A one molal solution of glucose, which does not ionize, reduces the freezing point by $1.86C°$ and raises the boiling point $0.512C°$.

QUESTIONS

A-1. What is the total number of hydrogen atoms in one mole of CH_3OH? (1) 6.0×10^{23} (2) 18×10^{23} (3) 24×10^{23} (4) 36×10^{23}

A-2. How many molecules are in 0.25 mole of O_2? (1) 12×10^{23} (2) 6.0×10^{23} (3) 3.0×10^{23} (4) 1.5×10^{23}

A-3. How do the freezing and boiling points of a sample of water change when 1 mole of NaCl is dissolved in it?
(1) The freezing point decreases and the boiling point increases.
(2) The freezing point increases and the boiling point increases.
(3) The freezing point decreases and the boiling point decreases.
(4) The freezing point increases and the boiling point decreases.

TABLE B

Standard Units

This table contains the names and symbols of several quantitative units commonly used in chemistry. It also lists some selected prefixes, with their symbols and exponential factor. The proper use of these standard units and prefixes is a very important skill in any science field.

QUESTIONS

B-1. The standard unit of energy is the   (1) kelvin (2) coulomb
(3) joule (4) pascal.
B-2. The quantity 2×10^6 volts could be expressed as (1) 2 millivolts
(2) 2000 V (3) 2MV (4) 2 microvolts
B-3. Atmospheric pressure, which is frequently expressed as mmHg, can also be expressed in terms of (1) mol (2) Pa (3) K (4) kg

TABLE C

Density and Boiling Points of Some Common Gases

In addition to density and boiling points, Table C contains.

1. Names and formulas of compounds.
2. Most of the diatomic gases.
3. The definition for STP
4. The relationship between atmospheres and torrs.
5. The relationship between the Kelvin and Celsius scales.

Densities can be used to determine molecular masses of gases, provided one is aware that one mole of any gas occupies a volume of 22.4 liters of STP.   Since density = mass/volume, the mass of one mole (the molecular mass) = density times 22.4 liters/mole.

Boiling points provide clues to the ease with which gases may be liquefied and relative diffusion rates.

QUESTIONS

C-1. Which Kelvin temperature is equal to –33°C?
  (1) –33 K   (2) 33 K   (3) 240 K   (4) 306K

C-2. Which gas will most closely resemble an ideal gas at STP?
  (1) SO_2 (2) NH_3 (3) Cl_2 (4) H_2

C-3. What is the density, in grams per liter, of N_2 gas at STP?
  (1) 28.0   (2) 14.0   (3) 1.25   (4) 0.800

C-4. Which 1.0 liter sample has the greatest mass at STP?
  (1) $H_2(g)$ (2) $CO_2(g)$ (3) $Cl_2(g)$ (4) $CH_4(g)$

C-5. Which gas has a greater density at STP than air at STP?
  (1) H_2 (2) NH_3 (3) Cl_2 (4) CH_4

C-6. Which liquid has a normal boiling point of 263 K?
  (1) $NO(\ell)$ (2) $CO(\ell)$ (3) $CO_2(\ell)$ (4) $SO_2(\ell)$

C-7. The density of NO gas in grams per liter at STP is approximately
  (1) 0.747   (2) 1.34   (3) 15.0   (4) 30.0

C-8. Which compound has the *lowest* normal boiling point?
  (1) HCl   (2) H_2S (3) NH_3 (4) CH_4

C-9. Which gas has the greatest density at STP?
  (1) nitrogen   (2) oxygen   (3) chlorine   (4) fluorine

C-10. Which formula represents a mixture?   (1) HCl(aq)   (2) HCl(g)
  (3) HCl(s)   (4) HCl(ℓ)

TABLE D

Solubility Curves

Each of the curves in this table represents the concentration, in grams solute per 100 grams water, of *saturated* solutions at various temperatures.

Some helpful hints:

1. Be certain to understand the definition of the *y* axis.
2. Questions often present names, rather than formulas, of compounds. If this represents a problem, be sure to take advantage of the information found in Tables E and F for the salts and Table C for the gases.
3. Note that most solids are more soluble at higher temperatures.
4. The gases in the table, HCl, NH_3 (ammonia), and SO_2, show a decrease in solubility with an increase in temperature. This behavior is typical of gases.

QUESTIONS

D-1. According to Reference Table D, approximately how many grams of $KClO_3$ are needed to saturate 100 grams of H_2O at 40°C?
(1) 6   (2) 16   (3) 38   (4) 47

D-2. How many grams of KNO_3 are needed to saturate 50. grams of water at 70.°C? (1) 30. g   (2) 65 g   (3) 130 g   (4) 160 g

D-3. Which quantity of salt will form a saturated solution of 100 grams of water at 45°C? (1) 30 g of KCl   (2) 35 g of NH_4Cl
(3) 60 g of KNO_3 (4) 110 g of $NaNO_3$

D-4. A solution containing 90. grams of KNO_3 per 100. grams of H_2O at 50.°C is considered to be (1) dilute and unsaturated (2) dilute and supersaturated (3) concentrated and unsaturated (4) concentrated and supersaturated

D-5. A solution contains 14 grams of KCl in 100. grams of water at 40.°C. What is the *minimum* amount of KCl that must be added to make this a saturated solution? (1) 14 g (2) 19 g (3) 25 g (4) 44 g

D-6. According to Reference Table D, how does a decrease in temperature from 40°C to 20° C affect the solubility of NH_3 and KCl? (1) The solubility of NH_3 decreases, and the solubility of KCl decreases   (2) The solubility of NH_3 decreases, and the solubility of KCl increases   (3) The solubility of NH_3 increases, and the solubility of KCl decreases. (4) The solubility of NH_3 increases, and the solubility of KCl increases.

TABLE E

Table of Solubilities in Water

It is important to be aware that salts resulting from reactions between solutions listed as "i" or "ss" are usually precipitated from solution. In other words, the quantities of these substances that do dissolve are extremely small. This provides evidence for predicting the formation of a precipitate whenever we know the identity of the ions in a mixture of ions.

Table E can be useful in naming compounds, in combination with Table F.

QUESTION

E-1. According to Reference Table E, which compound is most soluble in water? (1) $BaCO_3$ (2) $BaSO_4$ (3) $ZnCO_3$ (4) $ZnSO_4$

E-2. Which of the following substances is the *least* soluble in water at 25°C?   (1) NaCl (2) AgCl (3) NH_4Cl (4) KCl

E-3. When $PbI_2(s)$ is added to $Na_2CO_3(aq)$, a white precipitate is formed. According to Reference Table E, the white precipitate most likely is (1) KNO_3 (2) $PbCO_3$ (3) NaI (4) Na_2CO_3

TABLE F

Selected Polyatomic Ions

Table F has many applications in naming compounds and writing their formulas. Since the charge on each ion is presented, the table is also helpful in determining oxidation numbers. Notice that the convention in the notation for ionic charge is as follows:

1. When the charge is 1+ or 1−, only the sign is given, not the numeral.
2. For charges greater than one, the numeral is written first, then the charge. For example, 2+ or 3−.

Oxidation numbers are generally designated sign first, then numeral. For example, + 1, − 1, + 2, − 2, etc. (See Periodic Table of the Elements.) When naming acids of polyatomic ions, the rules are:

1. If the name of the polyatomic ion ends in –ite, the name of the acid associated with it ends in –ous. For example, nitr*ous* acid produces nitr*ite* ions.
2. If the name of the ion ends in –ate, the name of the acid ends in –ic. Nitr*ic* acid produces nitr*ate* ions.

QUESTIONS

F-1. The correct formula for the thiosulfate ion is (1) SO_3^{2-} (2) SO_4^{2-} (3) SCN^- (4) $S_2O_3^{2-}$

F-2. Which formula correctly represents mercury (I) chloride? (1) Hg_2Cl (2) $HgCl_2$ (3) Hg_2Cl_2 (4) Hg_2Cl_4

F-3. What is the name of the calcium salt of sulfuric acid? (1) calcium thiosulfate (2) calcium sulfate (3) calcium sulfide (3) calcium sulfite

F-4. What is the correct formula for sodium thiosulfate? (1) $Na_2S_2O_4$ (2) Na_2SO_3 (3) Na_2SO_4 (4) $Na_2S_2O_3$

TABLE G

Standard Energies of Formation of Compounds
at 1 atm and 298 K

The heat of formation (ΔH_f) is defined as the heat of reaction when *one mole* of a compound is formed from its elements. The sample equation in the table should remind the reader of this, since a mixed number coefficient is used for O_2 in order to meet the *one mole* restriction in the definition.

Some helpful hints:

1. If negative, the numerical value of ΔH_f^0 is the energy *released* in forming the compound. Therefore, it is the amount of energy *needed* to reverse the process. This means that large negative values of ΔH_f are associated with very stable compounds, with very little tendency to decompose spontaneously.

2. In order for compounds to be formed spontaneously, ΔG^0 must be negative.

3. ΔG is found from the equation $\Delta G = \Delta H - T\Delta S$ where T is the Kelvin temperature and S is the entropy of the system.

Table G is another table that provides assistance in naming compounds and writing formulas.

QUESTIONS

G-1. According to Reference Table G, which compound will form spontaneously from its elements? (1) nitrogen (IV) oxide (2) ethene (3) hydrogen iodide (4) potassium chloride

G-2. The change in free energy of a chemical reaction is represented by (1) ΔT (2) ΔS (3) ΔH (4) ΔG

G-3. Based on Reference Table *G*, which of the following compounds is the most stable? (1) CO (2) CO_2 (3) NO (4) NO_2

G-4. Based on Reference Table *G*, how many kilocalories of heat are given up when 0.500 mole of MgO(s) is formed from its elements? (1) 36.0 (2) 71.9 (3) 143.8 (4) 287.6

G-5. What is the free energy of formation of $H_2O(\ell)$ in kilocalories per mole at 1 atmosphere and 298 K? (1) –54.6 (2) –56.7 (3) –57.8 (4) –68.3

G-6. Based on Reference Table *G*, a compound which forms spontaneously from its elements is (1) NO (2) NO_2 (3) C_2H_4 (4) C_2H_6

TABLE H

Selected Radioisotopes

Table H is designed to be used in calculating rates, times, and/or masses involved during the decay of samples of the isotopes specified in a variety of conditions. *Nuclide* is a term used to refer to any isotope of an element. The particular isotope is indicated by the mass number, which is written as a superscript to the left of the symbol. ^{14}C is read, "carbon-fourteen," and refers to the isotope of carbon containing 8 neutrons. (Remember, mass number = protons + neutrons. All isotopes of carbon contain 6 protons. Thus, $14 - 6 = 8$ neutrons.

The key to utilizing this chart is the statement: "At the conclusion of each half-life, the mass of the sample is half of the mass it had at the beginning of that half-life." (See Unit 2.)

QUESTIONS

H-1. What is the number of hours required for potassium-42 to undergo 3 half-life periods? (1) 6.2 hours   (2) 12.4 hours   (3) 24.8 hours   (4) 37.2 hours

H-2. How many grams of a 32-gram sample of ^{32}P will remain after 71.5 days'?   (1) 1   (2) 2   (3) 8   (4) 4

TABLE I

Heats of Reaction at 1 atm and 298K

It is extremely important that the values in Table I not be confused with those in Table G, although they *are* related. Heats of reaction can be predicted on the basis of the Law of Conservation of Energy from Heats of Formation. (Heat of Formation of an element is assumed to be zero.)

Helpful Hints:

1. Where ΔH is *negative,* the reaction is *exothermic.*
2. Where ΔH is *positive,* the reaction is *endothermic.*
3. Exothermic reactions add heat to the environment and, therefore, raise the temperature of the reaction medium. Dissolving NaOH in water causes the temperature of the water to rise.
4. Endothermic reactions remove heat from the environment. Dissolving NH_4Cl in water lowers the temperature of the water.

Occasionally, the equations given in Table I provide information useful in answering other kinds of questions.

QUESTIONS

I-1. According to Reference Table I, the greatest amount of energy would be given up by the complete oxidation of 1 mole of
(1) $CH_4(g)$ (2) $C_3H_8(g)$ (3) $CH_3OH(\ell)$ (4) $C_6H_{12}O_6(s)$

I-2. According to Reference Table I, in which reaction do the products have a higher energy content than the reactants?

(1) $CH_4(g) + 2O_2(g) \rightarrow CO_2(g) + 2H_2O(\ell)$

(2) $CH_3OH(\ell) + \frac{3}{2}O_2(g) \rightarrow CO_2(g) + 2H_2O(\ell)$

(3) $NH_4Cl(s) \xrightarrow{H_2O} NH_4^+(aq) + Cl^-(aq)$

(4) $NaOH(s) \xrightarrow{H_2O} Na^+(aq) + OH^-(aq)$

TABLE J

Symbols Used in Nuclear Chemistry

In Table J, the columns are not labeled. The first column lists the names. The second column lists the identity of the particles, as well as their mass (the upper numeral) and charge (the lower numeral). The third column lists the conventional notation which is based on the name except for the electron (beta$^-$, or β^-), and the positron (beta$^+$ or β^-).

QUESTIONS

J-1. Which two particles have approximately the same mass?
(1) neutron and electron (2) neutron and deuteron (3) proton and neutron (4) proton and electron

J-2. Which nuclear emission moving through an electric field would be deflected toward the positive electrode?
(1) alpha particle (2) beta particle (3) gamma radiation (4) proton

J-3. Which particle is electrically neutral?   (1) proton (2) positron
(3) neutron   (4) electron

TABLE K

Ionization Energies and Electronegativities

Table K is an abbreviated form of the periodic table. The first ionization energy listed is the number of kilocalories of energy needed to remove one electron per atom from one mole of atoms. Therefore, elements that form positive ions by losing electrons have small ionization energies. Lowest values are found in the *lower left* regions of the table, among the most metallic elements. Highest values, as might be expected, are found among the inert or noble gases of Group 18. Among the elements other than Group 18, highest values are found among the halogens, the most non-metallic elements, in the *upper right* region of the table.

Table K also lists the electronegativity, which is a measure of the ease with which elements *gain* electrons to form negative ions. Highest values are found in the upper right region of the table and lowest values in the lower left. Differences in electronegativity account for the polar nature of bonds. Differences of 1.7 or more are associated with predominantly ionic bonds. Smaller differences indicate predominantly covalent bonds.

QUESTIONS

K-1. Which bond has the greatest degree of ionic character?
(1) H–Cl (2) I–Cl (3) Cl–Cl (4) K–Cl

K-2. Which of the following elements has the highest first ionization energy? (1) Li (2) Na (3) K (4) Rb

K-3. A molecule of ammonia (NH_3) contains (1) ionic bonds, only (2) covalent bonds, only (3) both covalent and ionic bonds (4) neither covalent nor ionic bonds

K-4. Given the same conditions, which of the following Group 17 elements has the *least* tendency to gain electrons?
(1) fluorine (2) iodine (3) bromine (4) chlorine

K-5. Which of the following atoms will lose an electron most readily?
(1) potassium (2) calcium (3) rubidium (4) strontium

TABLE L

Relative Strengths of Acids in Aqueous Solution at 1 atm and 298K

Table L is arranged so that the acids with the highest ionization constants (K_a) appear at the top of the list. These are referred to as the *strong* acids and they are associated with *weak* conjugate bases. Questions dealing with conjugate acids and bases can usually be answered by consulting this table. The equations found in Table L are often useful sources of information needed in many other test items, as well.

The term amphiprotic (or amphoteric) refers to substances that can act as acids (proton donors) or bases (proton acceptors). Table L reveals such substances by presenting them on both sides of the table. HSO_4^- is found as a base on the right side of the arrow in the fifth reaction from the top and as acid on the left side of the arrow in the seventh. It is an example of an amphoteric substance.

QUESTIONS

L-1. According to Reference Table L, which Brönsted acid has the strongest conjugate base? (1) H_2S (2) HS^- (3) NH_4^+ (4) NH_3

L-2. Based on Reference Table L, which 0.1 M solution is the best conductor of electricity? (1) HF(aq) (2) HI(aq) (3) HNO_2(aq) (4) H_2S(aq)

L-3. If HCl and H_2O react together in an acid-base reaction to form their Brönsted-Lowry conjugates, the products would be (1) HCl and H_3O^+ (2) Cl^- and OH^- (3) Cl_2 and H_2 (4) Cl^- and H_3O^+

L-4. According to Reference Table L, which substance is amphiprotic? (1) NO_2^- (2) S^{2-} (3) HSO_4^- (4) HF

L-5. Which is the conjugate base of water? (1) H^+ (2) OH^- (3) H_3O^+ (4) H_2O

L-6. According to Reference Table L, which of the following is the strongest Brönsted acid? (1) HNO_2 (2) H_2S (3) CH_3COOH (4) H_3PO_4

L-7. In the reaction $H_2O + H_2O \rightarrow H_3O^+ + OH^-$, the water is acting as (1) a proton acceptor, only (2) a proton donor, only (3) both a proton acceptor and donor (4) neither a proton acceptor nor donor

L-8. Which of the following acids ionizes to the *least* extent at 298 K? (1) HF (2) HNO_2 (3) H_2S (4) H_2O

TABLE M

Constants for various Equilibria at 1 atm and 298K

Table M presents equations and equbrium constants for certain reactions that are significant for different and special rasons. They can prove very useful in explaining effects of these reactions. The inonization constant for water ($K_w = 1.0 \times 10^{-14}$) provides useful information concerning acids and bases. Several characteristics of ammonia can be explained on the bases of the three reactions in which it appears. The equations are also useful in themselves.

The lower portion of the table deals with the solubility product constants (K_{sp}) of insoluble salts. The constants are helpful in calculating solubilities as well as predicting when precipitates will be formed based on concentrations of the ions that combine to produce them. The solubility of a salt is directly proportional to its K_{sp}; the greater the K_{sp}, the more soluble is the salt.

QUESTIONS

M-1. Based on Reference Table M, which of the following compounds is the most soluble? (1) AgCl (2) AgI (3) $PbCl_2$ (4) PbI_2

M-2. For the equilibrium reaction $AgCl(s) \rightleftarrows Ag^+(aq) + Cl^-(aq)$ at 25°C, the K_{sp} is equal to (1) 6.0×10^{-23} (2) 1.8×10^{-10} (3) 1.0×10^{-7} (4) 9.6×10^{-4}

M-3. According to Reference Table M, which compound is more soluble than $BaSO_4$ at 1 atmosphere and 298 K? (1) AgBr (2) $PbCl_2$ (3) AgI (4) ZnS

M-4. Based on Reference Table M, a saturated solution of which salt would be most dilute? (1) AgCl (2) $BaSO_4$ (3) ZnS (4) $PbCrO_4$

TABLE N

Standard Electrode Potentials

All the reactions in Table N are reduction reactions and involve the gain of electrons. They are referred to as "half-reations," because in order for them to take place, there must be an oxidation "half-reaction" occurring at the same time. Oxidation reactions involving the same ions are the reverse reactions of those shown in the table and are associated with voltages of the same magnitude but opposite sign. The *net potential* of the redox reaction can be calculated by adding the half cell potentials, making sure that the sign of the oxidation potential is opposite the one listed for the reverse reaction.

Given the two metal/metal ion half cells, the cathode must be the one listed higher on the table, since reduction always takes place at the cathode.

For example:

Given the electrochemical cell $Ag/Ag^+/ /Zn/Zn^+$, the Ag/Ag^+ is found in the table above the Zn/Zn^+ and the equation is:

$$Ag^+ + e^- \rightarrow Ag(s) +.80.$$

Since the Zn/Zn^+ is the oxidation, the equation must be reversed and the sign of the charge reversed:

$$Zn_{(s)} \rightarrow Zn^{2+} + 2e^- + .76$$

The net potential is $(+.80) + (+.76) = +1.56$ volts

In utilizing Table N, the following reminders are offered:

1. It is a *reduction* table in which potentials are measured by comparing the half cells with standard hydrogen electrodes.

2. Species most easily reduced appear highest in the table.

3. In redox reactions, the substance reduced is the oxidizing agent and the substance oxidized is the reducing agent.

4. Therefore, F_2 is the best oxidizing agent and Li is the best reducing agent.

5. Species more easily reduced than hydrogen have positive reduction potentials and those less easily reduced have negative values.

QUESTIONS

N-1. At 298 K, which metal will release $H_2(g)$ when reacted with HCl(aq)? (1) Au(s) (2) Zn(s) (3) Hg(ℓ) (4) Ag(s)

N-2. A standard hydrogen half-cell is connected to a standard silver half-cell by means of a wire and a salt bridge. The maximum standard potential $(E°)$ for the cell is (1) –0.80 volt (2) –1.60 volts (3) +0.80 volt (4) +1.60 volts

N-3. What is the reducing agent in the reaction

$$Pb + 2AgNO_3 \rightarrow Pb(NO_3)_2 + 2Ag?$$

(1) NO_3^- (2) Pb (3) Ag (4) Ag^+

N-4. According to Reference Table *N*, which species can reduce Fe^{3+} to Fe^{2+}? (1) Au(s) (2) Ag(s) (3) Br^- (4) I^-

N-5. Given the reaction: $Sn^{2+}(aq) + 2Fe^{3+}(aq) \rightleftarrows Sn^{4+}(aq) + 2Fe^{2+}(aq)$
The total number of moles of electrons lost by 1 mole of Sn^{2+} is
(1) 1 (2) 2 (3) 3 (4) 4

N-6. Hydrogen gas is produced when dilute hydrochloric acid is added to (1) copper (2) silver (3) gold (4) magnesium

N-7. Which atom-ion pair will react spontaneously under standard conditions? (1) $Mg + Li^+$ (2) $Mg + Ba^{2+}$ (3) $Mg + Ag^+$ (4) $Mg + Sr^{2+}$

N-8. According to Reference Table *N*, which halogen will react spontaneously with Au(s) to produce Au^{3+}? (1) Br_2 (2) F_2 (3) I_2 (4) Cl_2

TABLE O

Vapor Pressure of Water

The vapor pressure of water at various temperatures is especially useful in laboratory procedures involving the collection of gases by water displacement. When applied to Dalton's Law of Partial Pressures in conjunction with the total pressure in the collection vessel, the information in this table enables us to determine the actual pressure of the gas being collected. For example, if the barometric pressure in the lab is measured, and adjustments are made to insure that the pressure in the chamber is equal to that (by subtracting the vapor pressure of water at the temperature of the water in the collection vessel), the difference is the pressure of the gas collected. This value would then be used to calculate quantities of the collected gas in terms of mass or numbers of moles.

Given this Raw Data during collection of H_2 over water:

> Atmospheric Pressure = 754.0 torr
> Temperature of Water in Collection Vessel = 24°C
> Volume H_2 = 210.0

1. Vapor pressure of water at 24°C = 22.4 torr
2. Then the pressure due to the hydrogen would be

$$754.0 \text{ torr} - 22.4 \text{ torr} = 731.6 \text{ torr}$$

We could then conclude that the volume of hydrogen was 210.0 mL at 297K and 731.6 torr. These would be the values used for further calculations.

QUESTIONS

O-1. At what temperature will a sample of water boil when its vapor pressure is 30. torr? (1) 0°C (2) 29°C (3) 74°C (4) 100°C

O-2. As the atmospheric pressure increases, the temperature at which water in an open container will boil (1) decreases (2) increases (3) remains the same

O-3. A sample of pure water is boiling at 90.0°C. The vapor pressure of the water is closest to (1) 90.0 torr (2) 363 torr (3) 526 torr (4) 760. torr

O-4. As the temperature of a liquid increases, its vapor pressure (1) decreases (2) increases (3) remains the same

TABLE P

Radii of Atoms

Table P is a form of the periodic table in which the following data are given, in angstroms, for each element: Covalent Radius, Atomic Radius in Metals, and Van der Waals Radius.

Periodic Table of the Elements

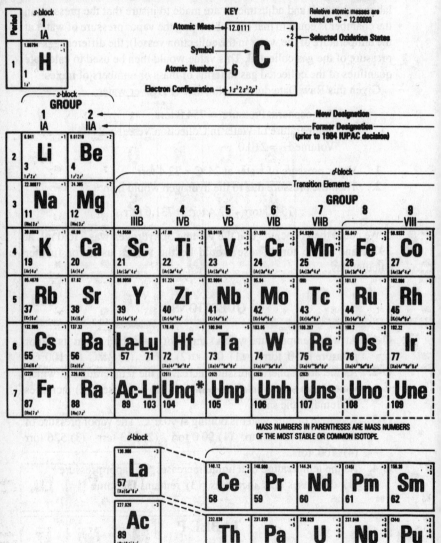

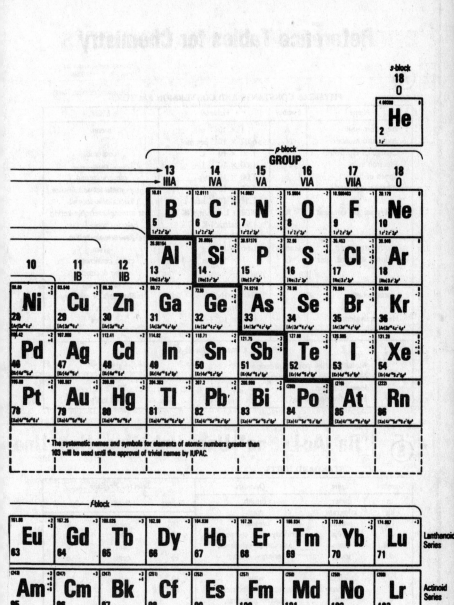

* The systematic names and symbols for elements of atomic numbers greater than 103 will be used until the approval of trivial names by IUPAC.

Reference Tables for Chemistry

(A)

PHYSICAL CONSTANTS AND CONVERSION FACTORS

Name	Symbol	Value(s)	Units
Angstrom unit	Å	1×10^{-10} m	meter
Avogadro number	N_A	6.02×10^{23} per mol	
Charge of electron	e	1.60×10^{-19} C	coulomb
Electron volt	eV	1.60×10^{-19} J	joule
Speed of light	c	3.00×10^8 m/s	meters/second
Planck's constant	h	6.63×10^{-34} J·s	joule-second
		1.58×10^{-37} kcal·s	kilocalorie-second
Universal gas constant	R	0.0821 L·atm/mol·K	liter-atmosphere/mole-kelvin
		1.98 cal/mol·K	calories/mole-kelvin
		8.31 J/mol·K	joules/mole-kelvin
Atomic mass unit	μ(amu)	1.66×10^{-24} g	gram
Volume standard, liter	L	1×10^3 cm^3 = 1 dm^3	cubic centimeters, cubic decimeter
Standard pressure, atmosphere	atm	101.3 kPa	kilopascals
		760 mmHg	millimeters of mercury
		760 torr	torr
Heat equivalent, kilocalorie	kcal	4.18×10^3 J	joules

Physical Constants for H$_2$O

Molal freezing point depression .... 1.86°C
Molal boiling point elevation............................. 0.52°C
Heat of fusion...................................... 79.72 cal/g
Heat of vaporization................................ 539.4 cal/g

(B)

STANDARD UNITS

Symbol	Name	Quantity	Selected Prefixes		
			Factor	Prefix	Symbol
m	meter	length			
kg	kilogram	mass			
Pa	pascal	pressure	10^6	mega	M
K	kelvin	thermodynamic temperature	10^3	kilo	k
mol	mole	amount of substance	10^{-1}	deci	d
J	joule	energy, work, quantity of heat	10^{-2}	centi	c
			10^{-3}	milli	m
s	second	time	10^{-6}	micro	μ
C	coulomb	quantity of electricity	10^{-9}	nano	n
V	volt	electric potential, potential difference			
L	liter	volume			

(Revised January 1987)

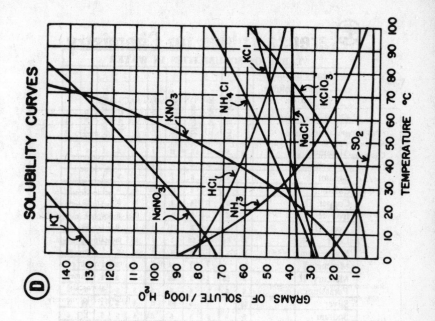

D SOLUBILITY CURVES

GRAMS OF SOLUTE/100g H_2O

TEMPERATURE °C

KI, NaNO₃, KNO₃, HCl, NH₄Cl, NH₃, NaCl, KClO₃, KCl, SO₂

C DENSITY AND BOILING POINTS OF SOME COMMON GASES

Name		Density grams/liter at STP*	Boiling Point (at 1 atm) K
Air	—	1.29	—
Ammonia	NH_3	0.771	240
Carbon dioxide	CO_2	1.98	195
Carbon monoxide	CO	1.25	82
Chlorine	Cl_2	3.21	238
Hydrogen	H_2	0.0899	20
Hydrogen chloride	HCl	1.64	188
Hydrogen sulfide	H_2S	1.54	212
Methane	CH_4	0.716	109
Nitrogen	N_2	1.25	77
Nitrogen (II) oxide	NO	1.34	121
Oxygen	O_2	1.43	90
Sulfur dioxide	SO_2	2.92	263

*STP is defined as 273K and 1 atm

E

TABLE OF SOLUBILITIES IN WATER

i — nearly insoluble
ss — slightly soluble
s — soluble
d — decomposes
n — not isolated

	acetate	bromide	carbonate	chloride	chromate	hydroxide	iodide	nitrate	phosphate	sulfate	sulfide
Aluminum	ss	s	n	s	n	i	s	s	i	s	d
Ammonium	s	s	s	s	s	s	s	s	s	s	s
Barium	s	s	i	s	i	s	s	s	i	i	d
Calcium	s	s	i	s	s	ss	s	s	i	ss	d
Copper II	s	s	i	s	i	i	n	s	i	s	i
Iron II	s	s	i	s	n	i	s	s	i	s	i
Iron III	s	s	n	s	i	i	n	s	i	ss	d
Lead	s	ss	i	ss	i	i	ss	s	i	i	i
Magnesium	s	s	i	s	s	i	s	s	i	s	d
Mercury I	ss	i	i	i	ss	n	i	s	i	ss	i
Mercury II	s	ss	i	s	ss	i	i	s	i	d	i
Potassium	s	s	s	s	s	s	s	s	s	s	s
Silver	ss	i	i	i	ss	n	i	s	i	ss	i
Sodium	s	s	s	s	s	s	s	s	s	s	s
Zinc	s	s	i	s	s	i	s	s	i	s	i

F

SELECTED POLYATOMIC IONS

Hg_2^{2+}	dimercury (I)	CrO_4^{2-}	chromate
NH_4^+	ammonium	$Cr_2O_7^{2-}$	dichromate
$C_2H_3O_2^-$	acetate	MnO_4^-	permanganate
CH_3COO^-		MnO_4^{2-}	manganate
CN^-	cyanide	NO_2^-	nitrite
CO_3^{2-}	carbonate	NO_3^-	nitrate
HCO_3^-	hydrogen carbonate	OH^-	hydroxide
		PO_4^{3-}	phosphate
$C_2O_4^{2-}$	oxalate	SCN^-	thiocyanate
ClO^-	hypochlorite	SO_3^{2-}	sulfite
ClO_2^-	chlorite	SO_4^{2-}	sulfate
ClO_3^-	chlorate	HSO_4^-	hydrogen sulfate
ClO_4^-	perchlorate	$S_2O_3^{2-}$	thiosulfate

G

STANDARD ENERGIES OF FORMATION OF COMPOUNDS AT 1 atm AND 298 K

Compound	Heat (Enthalpy) of Formation* kcal/mol ($\triangle H_f^o$)	Free Energy of Formation kcal/mol ($\triangle G_f^o$)
Aluminum oxide $Al_2O_3(s)$	-400.5	-378.2
Ammonia $NH_3(g)$	-11.0	-3.9
Barium sulfate $BaSO_4(s)$	-352.1	-325.6
Calcium hydroxide $Ca(OH)_2(s)$	-235.7	-214.8
Carbon dioxide $CO_2(g)$	-94.1	-94.3
Carbon monoxide $CO(g)$	-26.4	-32.8
Copper (II) sulfate $CuSO_4(s)$	-184.4	-158.2
Ethane $C_2H_6(g)$	-20.2	-7.9
Ethene (ethylene) $C_2H_4(g)$	12.5	16.3
Ethyne (acetylene) $C_2H_2(g)$	54.2	50.0
Hydrogen fluoride $HF(g)$	-64.8	-65.3
Hydrogen iodide $HI(g)$	6.3	0.4
Iodine chloride $ICl(g)$	4.3	-1.3
Lead (II) oxide $PbO(s)$	-51.5	-45.0
Magnesium oxide $MgO(s)$	-143.8	-136.1
Nitrogen (II) oxide $NO(g)$	21.6	20.7
Nitrogen (IV) oxide $NO_2(g)$	7.9	12.3
Potassium chloride $KCl(s)$	-104.4	-97.8
Sodium chloride $NaCl(s)$	-98.3	-91.8
Sulfur dioxide $SO_2(g)$	-70.9	-71.7
Water $H_2O(g)$	-57.8	-54.6
Water $H_2O(\ell)$	-68.3	-56.7

* Minus sign indicates an exothermic reaction.

Sample equations:

$$2Al(s) + \frac{3}{2} O_2(g) \rightarrow Al_2O_3(s) + 400.5 \text{ kcal}$$

$$2Al(s) + \frac{3}{2} O_2(g) \rightarrow Al_2O_3(s) \quad \triangle H = -400.5 \text{ kcal/mol}$$

(H)

SELECTED RADIOISOTOPES

Nuclide	Half-Life	Decay Mode
^{198}Au	2.69 d	β^-
^{14}C	5730 y	β^-
^{60}Co	5.26 y	β^-
^{137}Cs	30.23 y	β^-
^{220}Fr	27.5 s	α
^{3}H	12.26 y	β^-
^{131}I	8.07 d	β^-
^{37}K	1.23 s	β^+
^{42}K	12.4 h	β^-
^{85}Kr	10.76 y	β^-
^{85m}Kr*	4.39 h	γ
^{16}N	7.2 s	β^-
^{32}P	14.3 d	β^-
^{239}Pu	2.44×10^4 y	α
^{226}Ra	1600 y	α
^{222}Rn	3.82 d	α
^{90}Sr	28.1 y	β^-
^{99}Tc	2.13×10^5 y	β^-
^{99m}Tc*	6.01 h	γ
^{232}Th	1.4×10^{10} y	α
^{233}U	1.62×10^5 y	α
^{235}U	7.1×10^8 y	α
^{238}U	4.51×10^9 y	α

y=years; d=days; h=hours; s=seconds
*m = meta stable or excited state of the same nucleus. Gamma decay from such a state is called an isomeric transition (IT).
Nuclear isomers are different energy states of the same nucleus, each having a different measurable lifetime.

HEATS OF REACTION AT 1 atm and 298 K

Reaction	ΔH (kcal)
$CH_4(g) + 2O_2(g) \rightarrow CO_2(g) + 2H_2O(\ell)$	-212.8
$C_3H_8(g) + 5O_2(g) \rightarrow 3CO_2(g) + 4H_2O(\ell)$	-530.6
$CH_3OH(\ell) + \frac{3}{2}O_2(g) \rightarrow CO_2(g) + 2H_2O(\ell)$	-173.6
$C_6H_{12}O_6(s) + 6O_2(g) \rightarrow 6CO_2(g) + 6H_2O(\ell)$	-669.9
$CO(g) + \frac{1}{2}O_2(g) \rightarrow CO_2(g)$	-67.7
$C_8H_{18}(\ell) + \frac{25}{2}O_2(g) \rightarrow 8CO_2(g) + 9H_2O(\ell)$	-1302.7
$KNO_3(s) \xrightarrow{H_2O} K^+(aq) + NO_3^-(aq)$	$+8.3$
$NaOH(s) \xrightarrow{H_2O} Na^+(aq) + OH^-(aq)$	-10.6
$NH_4Cl(s) \xrightarrow{H_2O} NH_4^+(aq) + Cl^-(aq)$	$+3.5$
$NH_4NO_3(s) \xrightarrow{H_2O} NH_4^+(aq) + NO_3^-(aq)$	$+6.1$
$NaCl(s) \xrightarrow{H_2O} Na^+(aq) + Cl^-(aq)$	$+0.9$
$KClO_3(s) \xrightarrow{H_2O} K^+(aq) + ClO_3^-(aq)$	$+9.9$
$LiBr(s) \xrightarrow{H_2O} Li^+(aq) + Br^-(aq)$	-11.7
$H^+(aq) + OH^-(aq) \rightarrow H_2O(\ell)$	-13.8

(J)

SYMBOLS USED IN NUCLEAR CHEMISTRY		
alpha particle	^4_2He	α
beta particle (electron)	$^0_{-1}e$	β⁻
gamma radiation		γ
neutron	^1_0n	n
proton	^1_1H	p
deuteron	^2_1H	
triton	^3_1H	
positron	$^0_{+1}e$	β⁺

(K)

IONIZATION ENERGIES AND ELECTRONEGATIVITIES

1

313 ◄——— First Ionization Energy (kcal/mol of atoms)
H   ——— Electronegativity*
2.2

18

567
He

2	13	14	15	16	17	

125 Li 1.0	215 Be 1.5	191 B 2.0	260 C 2.6	336 N 3.1	314 O 3.5	402 F 4.0	497 Ne
119 Na 0.9	176 Mg 1.2	138 Al 1.5	188 Si 1.9	242 P 2.2	239 S 2.6	300 Cl 3.2	363 Ar
100 K 0.8	141 Ca 1.0	138 Ga 1.6	182 Ge 1.9	226 As 2.0	225 Se 2.5	273 Br 2.9	323 Kr
96 Rb 0.8	131 Sr 1.0	133 In 1.7	169 Sn 1.8	199 Sb 2.1	208 Te 2.3	241 I 2.7	280 Xe
90 Cs 0.7	120 Ba 0.9	141 Tl 1.8	171 Pb 1.8	168 Bi 1.9	194 Po 2.0	At 2.2	248 Rn
Fr 0.7	122 Ra 0.9						

* Arbitrary scale based on fluorine = 4.0

(L)

RELATIVE STRENGTHS OF ACIDS IN AQUEOUS SOLUTION AT 1 atm AND 298 K

Conjugate Pairs		
ACID	BASE	K_a
$HI = H^+ + I^-$		very large
$HBr = H^+ + Br^-$		very large
$HCl = H^+ + Cl^-$		very large
$HNO_3 = H^+ + NO_3^-$		very large
$H_2SO_4 = H^+ + HSO_4^-$		large
$H_2O + SO_2 = H^+ + HSO_3^-$		1.5×10^{-2}
$HSO_4^- = H^+ + SO_4^{2-}$		1.2×10^{-2}
$H_3PO_4 = H^+ + H_2PO_4^-$		7.5×10^{-3}
$Fe(H_2O)_6^{3+} = H^+ + Fe(H_2O)_5(OH)^{2+}$		8.9×10^{-4}
$HNO_2 = H^+ + NO_2^-$		4.6×10^{-4}
$HF = H^+ + F^-$		3.5×10^{-4}
$Cr(H_2O)_6^{3+} = H^+ + Cr(H_2O)_5(OH)^{2+}$		1.0×10^{-4}
$CH_3COOH = H^+ + CH_3COO^-$		1.8×10^{-5}
$Al(H_2O)_6^{3+} = H^+ + Al(H_2O)_5(OH)^{2+}$		1.1×10^{-5}
$H_2O + CO_2 = H^+ + HCO_3^-$		4.3×10^{-7}
$HSO_3^- = H^+ + SO_3^{2-}$		1.1×10^{-7}
$H_2S = H^+ + HS^-$		9.5×10^{-8}
$H_2PO_4^- = H^+ + HPO_4^{2-}$		6.2×10^{-8}
$NH_4^+ = H^+ + NH_3$		5.7×10^{-10}
$HCO_3^- = H^+ + CO_3^{2-}$		5.6×10^{-11}
$HPO_4^{2-} = H^+ + PO_4^{3-}$		2.2×10^{-13}
$HS^- = H^+ + S^{2-}$		1.3×10^{-14}
$H_2O = H^+ + OH^-$		1.0×10^{-14}
$OH^- = H^+ + O^{2-}$		$< 10^{-36}$
$NH_3 = H^+ + NH_2^-$		very small

Note: $H^+(aq) = H_3O^+$

Sample equation: $HI + H_2O = H_3O^+ + I^-$

(M)

CONSTANTS FOR VARIOUS EQUILIBRIA AT 1 atm AND 298 K

$H_2O(\ell) = H^+(aq) + OH^-(aq)$	$K_w = 1.0 \times 10^{-14}$
$H_2O(\ell) + H_2O(\ell) = H_3O^+(aq) + OH^-(aq)$	$K_w = 1.0 \times 10^{-14}$
$CH_3COO^-(aq) + H_2O(\ell) = CH_3COOH(aq) + OH^-(aq)$	$K_b = 5.6 \times 10^{-10}$
$Na^+F^-(aq) + H_2O(\ell) = Na^+(OH)^-(aq) + HF(aq)$	$K_b = 1.5 \times 10^{-11}$
$NH_3(aq) + H_2O(\ell) = NH_4^+(aq) + OH^-(aq)$	$K_b = 1.8 \times 10^{-5}$
$CO_3^{2-}(aq) + H_2O(\ell) = HCO_3^-(aq) + OH^-(aq)$	$K_b = 1.8 \times 10^{-4}$
$Ag(NH_3)_2^+(aq) = Ag^+(aq) + 2NH_3(aq)$	$K_{eq} = 8.9 \times 10^{-8}$
$N_2(g) + 3H_2(g) = 2NH_3(g)$	$K_{eq} = 6.7 \times 10^5$
$H_2(g) + I_2(g) = 2HI(g)$	$K_{eq} = 3.5 \times 10^{-1}$

Compound	K_{sp}	Compound	K_{sp}
$AgBr$	5.0×10^{-13}	Li_2CO_3	2.5×10^{-2}
$AgCl$	1.8×10^{-10}	$PbCl_2$	1.6×10^{-5}
Ag_2CrO_4	1.1×10^{-12}	$PbCO_3$	7.4×10^{-14}
AgI	8.3×10^{-17}	$PbCrO_4$	2.8×10^{-13}
$BaSO_4$	1.1×10^{-10}	PbI_2	7.1×10^{-9}
$CaSO_4$	9.1×10^{-6}	$ZnCO_3$	1.4×10^{-11}

N

STANDARD ELECTRODE POTENTIALS

Ionic Concentrations 1 M Water At 298 K, 1 atm

Half-Reaction	E^0 (volts)
$F_2(g) + 2e^- \rightarrow 2F^-$	+2.87
$8H^+ + MnO_4^- + 5e^- \rightarrow Mn^{2+} + 4H_2O$	+1.51
$Au^{3+} + 3e^- \rightarrow Au(s)$	+1.50
$Cl_2(g) + 2e^- \rightarrow 2Cl^-$	+1.36
$14H^+ + Cr_2O_7^{2-} + 6e^- \rightarrow 2Cr^{3+} + 7H_2O$	+1.23
$4H^+ + O_2(g) + 4e^- \rightarrow 2H_2O$	+1.23
$4H^+ + MnO_2(s) + 2e^- \rightarrow Mn^{2+} + 2H_2O$	+1.22
$Br_2(\ell) + 2e^- \rightarrow 2Br^-$	+1.09
$Hg^{2+} + 2e^- \rightarrow Hg(\ell)$	+0.85
$Ag^+ + e^- \rightarrow Ag(s)$	+0.80
$Hg_2^{2+} + 2e^- \rightarrow 2Hg(\ell)$	+0.80
$Fe^{3+} + e^- \rightarrow Fe^{2+}$	+0.77
$I_2(s) + 2e^- \rightarrow 2I^-$	+0.54
$Cu^+ + e^- \rightarrow Cu(s)$	+0.52
$Cu^{2+} + 2e^- \rightarrow Cu(s)$	+0.34
$4H^+ + SO_4^{2-} + 2e^- \rightarrow SO_2(aq) + 2H_2O$	+0.17
$Sn^{4+} + 2e^- \rightarrow Sn^{2+}$	+0.15
$2H^+ + 2e^- \rightarrow H_2(g)$	0.00
$Pb^{2+} + 2e^- \rightarrow Pb(s)$	−0.13
$Sn^{2+} + 2e^- \rightarrow Sn(s)$	−0.14
$Ni^{2+} + 2e^- \rightarrow Ni(s)$	−0.26
$Co^{2+} + 2e^- \rightarrow Co(s)$	−0.28
$Fe^{2+} + 2e^- \rightarrow Fe(s)$	−0.45
$Cr^{3+} + 3e^- \rightarrow Cr(s)$	−0.74
$Zn^{2+} + 2e^- \rightarrow Zn(s)$	−0.76
$2H_2O + 2e^- \rightarrow 2OH^- + H_2(g)$	−0.83
$Mn^{2+} + 2e^- \rightarrow Mn(s)$	−1.19
$Al^{3+} + 3e^- \rightarrow Al(s)$	−1.66
$Mg^{2+} + 2e^- \rightarrow Mg(s)$	−2.37
$Na^+ + e^- \rightarrow Na(s)$	−2.71
$Ca^{2+} + 2e^- \rightarrow Ca(s)$	−2.87
$Sr^{2+} + 2e^- \rightarrow Sr(s)$	−2.89
$Ba^{2+} + 2e^- \rightarrow Ba(s)$	−2.91
$Cs^+ + e^- \rightarrow Cs(s)$	−2.92
$K^+ + e^- \rightarrow K(s)$	−2.93
$Rb^+ + e^- \rightarrow Rb(s)$	−2.98
$Li^+ + e^- \rightarrow Li(s)$	−3.04

(O)

VAPOR PRESSURE OF WATER

°C	torr (mmHg)	°C	torr (mmHg)
0	4.6	26	25.2
5	6.5	27	26.7
10	9.2	28	28.3
15	12.8	29	30.0
16	13.6	30	31.8
17	14.5	40	55.3
18	15.5	50	92.5
19	16.5	60	149.4
20	17.5	70	233.7
21	18.7	80	355.1
22	19.8	90	525.8
23	21.1	100	760.0
24	22.4	105	906.1
25	23.8	110	1074.6

RADII OF ATOMS

KEY

Symbol	F
Covalent Radius, Å	0.64
Atomic Radius in Metals, Å	(−)
Van der Waals Radius, Å	1.35

A dash (−) indicates data are not available.

Each element below lists: **Covalent Radius (Å)**, **Atomic Radius in Metals (Å)**, **Van der Waals Radius (Å)**.

Element	Covalent	In Metals	Van der Waals
H	0.37	(−)	1.2
He	(−)	(−)	1.22
Li	1.23	1.52	(−)
Be	0.89	1.13	(−)
B	0.88	(−)	2.08
C	0.77	(−)	1.85
N	0.70	(−)	1.54
O	0.66	(−)	1.40
F	0.64	(−)	1.35
Ne	(−)	(−)	1.60
Na	1.57	1.54	2.31
Mg	1.36	1.60	(−)
Al	1.25	1.43	(−)
Si	1.17	(−)	2.0
P	1.10	(−)	1.90
S	1.04	(−)	1.85
Cl	0.99	(−)	1.81
Ar	(−)	(−)	1.91
K	2.03	2.27	2.31
Ca	1.74	1.97	(−)
Sc	1.44	1.61	(−)
Ti	1.32	1.45	(−)
V	1.22	1.32	(−)
Cr	1.17	1.25	(−)
Mn	1.17	1.24	(−)
Fe	1.17	1.24	(−)
Co	1.16	1.25	(−)
Ni	1.15	1.25	(−)
Cu	1.17	1.28	(−)
Zn	1.25	1.33	(−)
Ga	1.25	1.22	(−)
Ge	1.22	1.23	(−)
As	1.21	(−)	2.0
Se	1.17	(−)	2.0
Br	1.14	(−)	1.95
Kr	(−)	(−)	1.98
Rb	2.16	2.48	2.44
Sr	1.92	2.15	(−)
Y	1.62	1.81	(−)
Zr	1.45	1.60	(−)
Nb	1.34	1.43	(−)
Mo	1.29	1.36	(−)
Tc	(−)	1.36	(−)
Ru	1.24	1.33	(−)
Rh	1.25	1.35	(−)
Pd	1.28	1.38	(−)
Ag	1.34	1.44	(−)
Cd	1.41	1.49	(−)
In	1.50	1.63	(−)
Sn	1.40	1.41	(−)
Sb	1.41	(−)	2.2
Te	1.37	(−)	2.20
I	1.33	(−)	2.15
Xe	(−)	(−)	2.09
Cs	2.35	2.65	2.62
Ba	1.98	2.17	(−)
Hf	1.44	1.56	(−)
Ta	1.34	1.43	(−)
W	1.30	1.37	(−)
Re	1.28	1.37	(−)
Os	1.26	1.34	(−)
Ir	1.26	1.36	(−)
Pt	1.29	1.38	(−)
Au	1.34	1.44	(−)
Hg	1.44	1.60	(−)
Tl	1.55	1.70	(−)
Pb	1.54	1.75	(−)
Bi	1.52	1.55	(−)
Po	1.53	1.67	(−)
At	(−)	(−)	(−)
Rn	(−)	(−)	2.14
Fr	(−)	2.7	(−)
Ra	(−)	2.20	(−)

Lanthanides (La–Lu)

Element	Covalent	In Metals	Van der Waals
La	1.69	1.88	(−)
Ce	1.65	1.83	(−)
Pr	1.65	1.83	(−)
Nd	1.64	1.82	(−)
Pm	(−)	1.81	(−)
Sm	1.66	1.80	(−)
Eu	1.85	2.04	(−)
Gd	1.61	1.80	(−)
Tb	1.59	1.78	(−)
Dy	1.59	1.77	(−)
Ho	1.58	1.77	(−)
Er	1.57	1.76	(−)
Tm	1.56	1.75	(−)
Yb	1.70	1.94	(−)
Lu	1.56	1.73	(−)

Actinides (Ac–Lr)

Element	Covalent	In Metals	Van der Waals
Ac	(−)	1.88	(−)
Th	1.80	1.80	(−)
Pa	1.61	1.61	(−)
U	(−)	1.39	(−)
Np	(−)	1.31	(−)
Pu	(−)	1.51	(−)
Am	(−)	1.84	(−)
Cm	(−)	(−)	(−)
Bk	(−)	(−)	(−)
Cf	(−)	(−)	(−)
Es	(−)	(−)	(−)
Fm	(−)	(−)	(−)
Md	(−)	(−)	(−)
No	(−)	(−)	(−)
Lr	(−)	(−)	(−)

GLOSSARY

absolute zero: The coldest possible temperature; OK or $-273°C$.

acid: A substance which contains hydrogen and yields hydrogen ions in water; a proton donor

activated complex: The temporary, unstable, intermediate union of reactants.

activation energy: The minimum amount of energy needed to produce an activated complex.

addition reaction: A reaction in organic chemistry in which atoms are added to a compound at the site of double or triple bonds.

alcohols: A family of organic compounds. The molecules in this family consist of a hydrocarbon radical combined with one or more hydroxyl $(-OH)$ groups. Their IUPAC name ends in -ol.

aldehydes: A family of organic compounds. The molecules in this family consist of a hydrocarbon radical combined with a terminal $-CHO$ group. Their IUPAC name ends in -al.

aliphatics: Families of hydrocarbons with open carbon chains.

alkali metal: A member of Group IA of the Periodic Table. An active metal having an oxidation value of $+1$ in compounds.

alkaline metal: Also called an Alkaline Earth metal. A member of Group IIA of the Periodic Table having an oxidation value of $+2$ in compounds.

alkanes: The family of saturated hydrocarbons, the molecules of which have the general formula C_nH_{2n+2} and contain only single covalent carbon-to-carbon bonds.

alkenes: A family of unsaturated hydrocarbons. The molecules of this family have the general formula C_nH_{2n} and contain one double covalent carbon-to-carbon bond.

alkyl group: A hydrocarbon radical having the general formula C_nH_{2n+1}.

alkynes: A family of unsaturated hydrocarbons the molecules of which have the general formula C_nH_{2n-2}. These compounds contain one triple covalent carbon-to-carbon bond.

alloy: A homogeneous mixture containing two or more metals.

alpha decay: A transmutation in which a helium nucleus is emitted and the resulting daughter nucleus has an atomic number reduced by 2 and an atomic mass reduced by 4.

alpha particle: A helium nucleus.

amine: A type of organic compound the molecules of which consist of a hydrocarbon radical with a functional $-NH_2$ attached.

amphiprotic: A substance that can act either as a proton donor (acid) or a proton acceptor (base).

amphoteric: Property of hydroxides that also act as acids; commonly the hydroxides of aluminum and zinc.

analysis: The decomposition of a compound into simpler substances.

anhydrous: Term applied to a hydrate after the water of hydration has been removed.

anion: A negatively charged ion.

anode: The site of oxidation in an electrochemical cell. It is considered to be negatively charged in an electrochemical cell and positively charged in an electrolytic cell.

aqueous: Term applied to a solution in which water is the solvent. It is indicated by (aq), e.g., NaCl (aq).

aromatics: A family of organic compounds. The molecules of this family contain one or more benzene rings.

Arrhenius acid: A substance that yields the hydronium ion in water solution.

Arrhenius base: A substance that contains the hydroxide ion in water solution.

atomic mass: The average mass of the naturally occurring isotopes of an element.

atomic mass units: A relative mass scale with a basic unit $\frac{1}{12}$ the mass of carbon-12.

atomic number: The number of protons in the nucleus of an atom; sometimes designated by the letter Z.

atomic radius: A measure of one-half the distance between two nuclei of an element in the solid phase.

Avogadro's hypothesis: Equal volumes of gases, when measured under the same conditions of temperature, contain the same number of molecules.

Avogadro's number: The number of particles in a mole of a compound— 6.02×10^{23}.

base: A proton acceptor (Brönsted); a substance that donates a pair of electrons in chemical action (Lewis).

basic anhydride: An axide of a metal that can react with water to form a base.

beta particles (rays): High-velocity streams of electrons. Term usually applied to electrons emitted by radioactive atoms or to artificially accelerated electrons (as in the betatron).

binary compound: A compound made up of only two elements.

binding energy: The energy equivalent to the mass defect of an atomic nucleus.

boiling: Turbulence in a liquid caused by the rapid formation of bubbles of vapor at the boiling point of the liquid. See **boiling point**.

boiling point: The temperature at which the vapor pressure of a liquid is equal to the pressure exerted on the liquid. When the atmospheric pressure is 760 mm of mercury, the boiling point of water is, by definition, 100°C, or 373K.

bond strength: A measure of the amount of energy that must be supplied to break a chemical bond.

calorie: A unit of heat. The heat required to raise the temperature of one gram of water one Celsius degree.

calorimeter: An apparatus for measuring the quantity of heat liberated or absorbed during a reaction.

carbohydrate: A polyhydroxy derivative of a saturated aldehyde or ketone.

catalysis: The change in the rate of a reaction by the presence of a substance which is unchanged at the end of the reaction.

catalytic agent: A substance which alters the rate of a chemical change and which remains unchanged at the end of the reaction, e.g., manganese dioxide in the preparation of oxygen.

cathode: (1) The negative electrode of an electrolytic cell. (2) The electrode at which reduction takes place.

cathode rays: Streams of electrons that emanate from the cathode electrode of a discharge tube.

cation: A positive ion. A cation is attracted to the cathode in an electrolysis, in which the cathode is the negative electrode.

Celsius scale: The temperature scale on which the temperature of freezing water is given the value $0°C$, and the temperature of boiling water is $100°C$.

chain reaction: A series of reactions in which each reaction is initiated by the energy produced in the preceding reaction.

chemical bond: The linkage between atoms due to the attraction of opposite charges, to the magnetic attractions of shared electrons, or to a combination of such forces.

chemical change: A change in the composition of substances with accompanying changes in properties.

chemical equilibrium: A condition in which two chemical changes exactly oppose each other. Equilibrium is a dynamic condition in which concentrations do not change, and the rate of the forward reaction is equal to the rate of the reverse reaction.

chemical property: A characteristic of a substance that is observed only when the substance undergoes chemical change.

chemistry: The study of the composition, structure, and properties of matter, the changes which matter undergoes, and the energy accompanying these changes.

coefficients: The numbers preceding the formulas in chemical equations, indicating the smallest number of molecules of the substance that may take part in the reaction.

colloidal state: That state of matter in which the particle size lies between 10^{-7} cm and 10^{-4} cm in diameter. The colloidal particle is larger in size than ordinary molecules, but smaller in size than particles which can be seen by the ordinary microscope.

combustion: Any chemical action producing noticeable light and heat.

common ion effect: A displacement of equilibrium brought about by increasing the concentration of one of the ions involved.

complex ions: Charged particles which contain more than one atom. Most complex-ion theory is applied to those complex ions which break down, in chemical action, to smaller units.

compound: A substance of definite composition which may be decomposed into two or more elements by chemical change.

concentrated solution: A solution containing a relatively large amount of solute.

concentration: The quantity of a substance in a given volume, expressed as molarity, molality, normality, percentage, density, etc.

condensation: The process whereby a gas or vapor changes to a liquid.

conjugates: The term applied to acid-base pairs which may be formed reversibly from one another in a protolysis reaction. Conjugates differ in composition only by a hydrogen ion. Many pairs are used in buffer mixtures. Examples: NH_3 and NH_4^+; H_2O and OH^-; H_2O and H_3O^+; CH_3COOH and CH_3COO^-.

coordinate covalence: Covalent bonding of two atoms in which one of the atoms furnishes both of the shared electrons.

covalent bond: A bond indicating a pair of shared electrons.

crystal: A solid with a definite shape, made up of plane faces. The shape is due to the atoms or molecules arranged in a definite repeated pattern.

decay: The spontaneous transmutation of a radioactive nucleus.

decomposition: Applied to a chemical change in which a substance breaks down to form two or more simpler substances.

density: Mass per unit volume, e.g., grams per cubic centimeter.

destructive distillation: A process in which a substance (usually organic) is heated in the absence of air until decomposition takes place. The products of decomposition (volatile matter) are condensed and collected.

deuterium: An isotope of hydrogen of mass 2.

deuteron: The nucleus of the deuterium atom.

diene: Aliphatic hydrocarbon containing 2 double bonds.

ductility: That property of a substance which permits its being drawn into wire.

dihydroxy alcohol: An alcohol whose molecules contain two hydroxy groups per molecule. Antifreeze (ethylene glycol) is such a substance.

dilute: Term applied to a solution that contains a relatively small amount of solute. A process by which the concentration of a solution is reduced.

dipole: A molecule that has an uneven charge distribution. Asymmetrical molecules that contain polar covalent bonds are dipoles.

dissociation: The separation of the ions of an ionic compound, especially during the process of dissolving.

distillate: The product produced by condensation of the vapors produced during distillation.

distillation: The process by which a substance is boiled and the vapors are condensed and recovered.

double bond: The sharing of two pairs of bonding electrons between two atoms.

electrochemical cell: A cell in which the redox reaction is conducted in such a way that the electrons travel through a wire between the substance being oxidized and the substance being reduced.

electrolysis: A chemical reaction that takes place when an electric current is applied to a substance.

electrolyte: A substance whose water solution conducts an electric current.

electron: A fundamental particle of matter having a negative electric charge.

electron-dot symbols: Symbols that contain the symbol of the element and indicate the number of valence electrons.

electronegativity: A measure of the attraction of a nucleus for the electrons in a covalent bond.

electroplating: The process of layering a metal onto a surface in an electrolytic cell.

element: A substance that cannot be decomposed by ordinary chemical means.

empirical formula: A formula showing the simplest ratio of the elements in a chemical compound.

endothermic reaction: A reaction in which the products contain more potential energy than the reactants. ΔH for an endothermic reaction is positive.

end point: That point of a titration when an indicator shows that equivalent amounts of reactants have reacted.

energy: Often defined as the ability to do work. The total amount of energy in a reaction must remain the same. However, it may be changed from one type of energy to another.

enthalpy: A measure of the potential energy of a substance.

entropy: A measure of the amount of randomness of the particles of a substance.

equilibrium: A dynamic chemical condition in which opposing reactions are proceeding at equal rates, producing an apparent constant condition.

esterification: The reaction of an acid and an alcohol to produce water and an ester.

evaporation: The changing of a substance from the liquid to the gaseous phase by the absorption of heat.

excited atom: The state of an atom when an electron moves to a higher energy level leaving a lower energy level vacant.

exothermic reaction: A reaction in which the products contain less energy than the reactants.

families: The vertical groupings of chemically similar elements in the Periodic Table.

fermentation: The production of ethanol and carbon dioxide by the action of enzymes on an organic compound.

filtration: A method used to separate solids from liquids.

fission: A nuclear reaction in which large nuclei are split into smaller nuclear fragments.

formula mass: The sum of all of the atomic masses in a formula; primarily used to describe the mass of ionic substances.

fractional distillation: The separation of different liquids in a mixture by using the different boiling points of the components.

free energy: A measure of the tendency of a reaction to proceed spontaneously. It is represented by ΔG in the Gibbs equation:
$$\Delta G = \Delta H - (T\Delta S).$$

freezing point: The temperature at which both the solid and liquid phases of a substance can exist in equilibrium.

freezing point depression: The lowering of the normal freezing point of a liquid by the addition of solute. One mole of particles lowers the freezing point of 1L of water by $1.86°C$.

fuctional group: An atom or group of atoms responsible for specific properties and characteristics of organic compounds.

fusion: 1. The change of a substance from the solid to the liquid phase. 2. A nuclear reaction in which light nuclei combine to form a heavier nucleus.

gamma rays: High energy X-rays emitted from the nucleus of a radioactive element.

gas density: The mass of a liter of gas expressed in grams per liter.

gas phase: The phase of matter which has neither definite volume nor shape.

Gay-Lussac's Law: The volumes of combining gases are in small whole number ratios.

gram-atomic mass: The gram amount of an element numerically equal to the atomic mass of the element.

gram-molecular mass: The mass of a substance in grams numerically equal to the molecular mass of a compound.

ground state: The condition of an atom in which the electrons occupy the lowest available energy levels.

groups: The vertical columns of the periodic table; also called families.

half-life: The length of time needed for one-half of a given radioactive substance to undergo decay.

half reaction: Either the oxidation or reduction portion of a redox reaction.

halogen: A member of Group VIIA of the Periodic Table.

halogenation: The placing of a halogen on a carbon chain.

heat: The flow of energy between objects of unequal temperature.

heat of condensation: The amount of heat released as a unit of mass of a substance changes from a vapor to a liquid.

heat of formation: Amount of heat gained or lost during the formation of one mole of a compound from its elements.

heat of fusion: The amount of heat needed to change a unit mass of a substance from solid to liquid phase.

heat of reaction: The amount of heat released or absorbed in a reaction.

heat of solidification: The amount of heat released as a unit mass of a substance changes from the liquid to the solid phase.

heat of vaporization: The amount of heat needed to evaporate a unit mass of a liquid at its boiling point.

heterogeneous: Consisting of different ingredients.

homogeneous: Having similar properties throughout.

hydrocarbon: An organic compound which contains only carbon and hydrogen.

hydrogen bond: A bond formed between a covalently bonded hydrogen atom and an atom with a high electronegativity value.

hydrogenation: The addition of hydrogen to a substance.

ideal gas: A theoretical gas which occupies no volume and whose particles have no attraction for each other.

ion: A charged atom or group of charged atoms.

ionic bond: A bond formed by the exchange of an electron between two atoms.

ionization constant: The equilibrium constant of a reversible reaction in which ions are produced from molecules.

ionization energy: The amount of energy needed to remove an electron from a neutral gaseous atom.

isomers: Compounds that have the same molecular formula, but different structural formulas.

isotopes: Nuclei that have the same number of protons but different numbers of neutrons and hence different atomic masses.

IUPAC: International Union of Pure and Applied Chemistry.

IUPAC nomenclature: A system of naming organic compounds as approved by the IUPAC.

Kelvin scale: Also called the absolute temperature scale. The zero point is the coldest possible temperature. One Kelvin degree is equivalent to one Celsius degree. A Celsius reading can be converted to Kelvin by adding 273 to the Celsius reading.

kernel: The nucleus and electrons of an atom, except the valence electrons.

ketones: A family of organic compounds whose molecules contain the carbonyl group as the functional group. Propanone (acetone) is the most common member.

kinetic energy: Energy of motion.

kinetic theory: A theory that explains the behavior of gases in terms of the motion of their molecules.

liquid: A phase of matter having a definite volume but taking the shape of its container.

litmus: An indicator that is red in acidic and blue in basic solutions.

macromolecules: Molecules formed by covalent network bonding that are large enough to be seen by the eye.

mass defect: The amount of matter that was converted into energy as protons and neutrons combined to form nuclei.

mass number: The sum of the protons and neutrons in a nucleus.

melting point: The temperature at which a substance melts and can coexist with the liquid phase of that substance. It is the same as **freezing point**, and involves the addition of potential energy to the substance without a temperature change.

meniscus: The curved surface of a liquid in a container. Instruments are calibrated so that volume readings are taken at the bottom of the meniscus.

metals: Atoms that lose electrons in chemical reactions to become positive ions.

metallic bond: The force that holds metallic atoms together in the solid or liquid phase. It is due to attraction between valence electrons and the positive kernels.

metalloid: Members of the Periodic Table that have both metallic and nonmetallic characteristics.

miscible: Term applied to substances that can be mixed to form a solution.

mixture: A substance not having definite proportions and containing two or more components that are not chemically combined.

moderator: A substance used to slow down the neutrons produced in a nuclear reactor.

molal boiling point constant: The number of degrees that 1 mole of solute particles will elevate the boiling point of 1,000g of solvent. The constant for 1L of water is 0.52°C.

molal freezing point constant: The number of degrees that 1 mole of solute particles will depress the freezing point of 1,000g of solvent. The constant for 1L of water is 1.86°C.

molality: The concentration of a solution expressed as the number of moles of solute per 1,000g of solvent.

molar volume: the volume occupied by 1 mole of a gas. This volume is 22.4L at STP; it contains 1 mole of molecules.

molarity: The concentration of a solution expressed as the number of moles of solute per liter of solution.

mole: 6.02×10^{23} particles. This number of particles can be obtained by taking the molecular mass of a substance in grams.

molecular formula: A formula that indicates the number of atoms that are present in the smallest particle that has the chemical properties of the substance.

molecular mass: The sum of the atomic masses of the atoms in a molecular formula.

molecule: The smallest unit of a substance that has the chemical properties of the substance; a discrete particle formed by covalently bonded atoms.

monohydroxy alcohol: An alcohol whose molecules contain only one hydroxyl group.

network solid: A solid formed by covalent bonding in which the bonding extends throughout the entire visible sample.

neutralization: A reaction between an Arrhenius acid and an Arrhenius base to produce a salt and water.

neutron: A neutral nuclear particle having a mass of 1 amu.

noble gas: Also called an **inert gas**. A member of Col O of the Periodic Table. These gases have filled valence electron shells and generally do not enter chemical reactions.

nonelectrolyte: An aqueous solution that does not conduct an electric current.

nonmetals: Elements whose valence electron shells are almost complete.

nonpolar bonds: Covalent bonds in which the two atoms have an equal share of the bonding electrons as measured by their electronegativity values.

nonpolar molecules: Molecules that have a symmetrical shape and a symmetrical charge distribution.

normal boiling point: The boiling point of a liquid at standard pressure.

nuclear fuel: The element used as the source of energy in a nuclear reactor. Often U-235.

nucleons: Particles found in the nucleus. Protons and neutrons are tne most commonly identified.

nucleus: The small, dense center of an atom that contains almost all of the mass of the atom in the form of protons and neutrons.

number of sublevels: The number of sublevels at a given principal quantum level is the same as the principal quantum level (n).

octet: A stable configuration of 8 valence electrons.

orbital: The area or space of an atom where an electron of a particular energy content is most likely to be found.

organic acids: Organic compounds the molecules of which contain the carboxyl (−COOH) group as their functional group.

organic chemistry: The chemistry of carbon compounds, particularly hydrocarbons and their derivatives.

ore: A mineral that can be used to produce a metal.

oxidation: The process by which a particle loses an electron or appears to lose an electron as indicated by a gain in oxidation number.

oxidation numbers: Values assigned to particles for the purpose of identifying oxidation and reduction processes.

oxidizing agents: Particles that cause other particles to be oxidized by accepting electrons from them while themselves being reduced.

particle accelerators: Devices that accelerate charged particles by magnetic or electric fields.

period: A horizontal sequence of elements on the Periodic Table that begins with an alkali metal and ends with noble (inert) gas. The first period of the Table begins with hydrogen and ends with helium.

Periodic Law: The properties of the elements are a periodic function of their atomic numbers.

peroxide: A binary compound that contains more oxygen atoms than are normally expected. Oxygen is assigned the oxidation value of −1 in such compounds.

pH: A method of expressing the acidity or basicity of a substance on a scale from 1 to 14. It is the negative of the logarithm of the hydrogen ion concentration.

phase: refers to the gas, liquid, or solid condition of matter.

phenolphthalein: An indicator that is colorless in acidic solutions and pink in basic solutions.

pOH: The negative of the logarithm of the hydroxide ion concentration.

polar bond: A bond in which the electron pair is shared unequally by the two atoms, resulting in a dipole. The element with the higher electronegativity value is assigned the negative portion of the dipole.

polar molecule: A molecule containing polar bonds with an asymmetric shape and therefore an asymmetric charge distribution.

polyatomic ion: Two or more atoms that are chemically combined and possess a net electric charge, also called a **radical**.

polymer: A compound with a high molecular mass that consists of many smaller subunits (monomers) that have been bonded together.

polymerization: The process of forming molecules of high molecular mass by the joining of smaller molecules into a chain.

positron: A fundamental particle with a mass identical to that of an electron, but having a positive charge.

potential energy: Often called stored energy. A particle has potential energy because of its position, phase, or composition.

precipitate: A solid that is formed when two liquids are mixed.

primary alcohols: Alcohol molecules in which the hydroxy (OH) group is attached to a primary carbon.

primary carbon: A carbon in a chain of carbons that is directly attached to one other carbon. These carbons are found at the ends of chains or branches.

proton: A particle found in the nucleus having a charge of $+1$ and having a mass of 1 amu.

radiation: Energy emitted from an object.

radioactive dating: A method of determining the age of an object by the use of the half-lives of radioactive elements in the sample.

radioactivity: The spontaneous release of energy by a nucleus.

radioisotope: A radioactive isotope of an element.

reactant: One of the substances consumed in a chemical reaction, a starting substance.

redox: Term used to describe the process in which oxidation and reduction take place.

reducing agent: A particle that causes another particle to be reduced while losing electrons, or being oxidized.

reduction: The process by which a particle gains electrons, as identified by a decrease in oxidation number.

reversible reaction: A reaction in which the products can reform into the reactants.

salt: An ionic substance consisting of a positive metallic ion and a negative ion other than the hydroxide ion.

salt bridge: A passageway for the movement of ions in an electrochemical cell.

saponification: The reaction of a base plus an ester to produce an alcohol and a soap.

saturated hydrocarbon: A hydrocarbon molecule containing only single covalent carbon-to-carbon bonds.

saturated solution: A solution in which as much solute has been dissolved as is possible for the given temperature.

secondary alcohol: An alcohol in which the hydroxy group is attached to a secondary carbon.

secondary carbon: A carbon in an organic compound that is directly attached to two other carbon atoms.

single bonds: Covalent bonds between atoms in which one pair of electrons is shared between the atoms.

solid: The phase of matter whose particles have a definite crystalline arrangement. Solids have both definite volume and a definite shape.

solubility curve: A curve showing the solubility of a solute as a function of temperature.

solubility product constant: The form of the equilibrium expression that applies to the equilibrium between an ionic solid and its ions. The larger the value, the more soluble the salt.

solute: The dissolved portion of a solution; the substance present in lesser amount.

solution: A homogeneous mixture of solute in solvent.

solvent: The part of a solution in which the solute is dissolved; the substance present in the greater amount.

spectrum: The series of lines of radiant energy produced as electrons return from higher to lower energy levels.

spontaneous reaction: A reaction that once begun will continue until one of the reactants has been consumed.

standard conditions: 760 torr and O°C. Referred to as **STP.**

standard electrode potential: The voltage of a half-cell in combination with a standard hydrogen cell.

standard solution: A solution of known concentration.

STP: Standard temperature and pressure.

Stock system: A system of nomenclature in which a Roman numeral is used to show the charge of any metallic ion that can carry different charges.

stoichiometry: The study of the quantitative aspects of formulas and equations.

strong acids: Acids that are highly ionized in solution. The degree of ionization is indicated by the magnitude of the ionization constant.

strong bases: Bases that are highly ionized in solution.

sublevels: Divisions of the principal atomic energy levels. The number of possible sublevels is the same as the principal quantum number.

sublimation: A change between the solid and gaseous phases without a noticeable liquid phase.

substance: any variety of matter, all specimens of which have identical properties and composition.

supersaturated solution: A solution that contains more dissolved solute than would be present in a saturated solution at the same temperature.

symbol: An upper case letter or an upper case letter plus a lower case letter used to represent an atom of an element or one mole of atoms of that element.

temperature: A measure of the average kinetic energy of the particles of a substance.

ternary acid: An acid containing three different elements per molecule.

tertiary alcohol: An alcohol molecule whose hydroxyl group is located on a tertiary carbon.

tertiary carbon: A carbon atom that is directly attached to three carbon atoms.

tincture: A solution in which the solvent is ethanol.

titration: A process in which a solution of known concentration is used to determine the concentration of another solution.

torr: A unit of pressure. Each torr is equivalent to 1 mm Hg. Standard pressure is 760 torr.

tracer: A radioisotope used to follow the path of a chemical reaction.

transition element: An element with an incomplete sublevel other than its valence sublevel.

transmutation: The conversion of atoms of one element into a different element.

transuranic element: An element with an atomic number greater than 92.

trihydroxy alcohol: An alcohol molecule with three hydroxy groups, such as 1,2,3, propanetriol (glycerol or glycerine).

triple bond: The sharing of three pairs of electrons between two atoms.

tritium: The isotope of hydrogen whose nuclei contain one proton and two neutrons and have a mass of 3 amu.

unsaturated hydrocarbon: A hydrocarbon molecule that contains at least one double or triple bond.

unsaturated solution: A solution in which more solute can be dissolved at a given temperature.

valence electrons: The s and p electrons at the highest principal quantum level, the outermost electrons of an atom.

van der Waals forces: Weak attractive forces between molecules in the solid and liquid phases. They are the result of temporary dipoles in molecules caused by the random, asymmetric motion of electrons.

vapor: The gaseous phase of a substance that is normally a solid or liquid at room temperature.

vapor pressure: The pressure exerted by the vapors of a liquid or a solid.

volt: A unit of electrical potential.

water of hydration: Also known as water of crystallization; the number of water molecules chemically attached to a particle of the substance in the solid state.

weak acid: An acid that is only slightly ionized. It will have a small K_a or K_{ion} value.

weak base: A base that is only ionized slightly. It will have a small K_{ion} value.

weak electrolyte: A substance whose water solution is a poor conductor of electricity.

INDEX

Regents Examinations

CHEMISTRY - June 20, 1996

Part I

Answer all 56 questions in this part. [65]

Directions (1-56): For *each* statement or question, select the word or expression that, of those given, best completes the statement or answers the question. Record your answer on the separate answer paper.

1. What is the vapor pressure of a liquid at its normal boiling temperature? (1) 1 atm (2) 2 atm (3) 273 atm (4) 760 atm

2. A sealed container has 1 mole of helium and 2 moles of nitrogen at 30°C. When the total pressure of the mixture is 600 torr, what is the partial pressure of the nitrogen? (1) 100 torr (2) 200 torr (3) 400 torr (4) 600 torr

3. Solid X is placed in contact with solid Y. Heat will flow spontaneously from X to Y when (1) X is 20°C and Y is 20°C (2) X is 10°C and Y is 5°C (3) X is −25°C and Y is −10°C (4) X is 25°C and Y is 30°C

4. Which graph represents the relationship between volume and Kelvin temperature for an ideal gas at constant pressure?

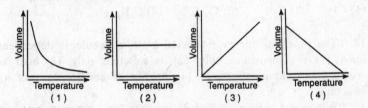

5. An example of a binary compound is (1) potassium chloride (2) ammonium chloride (3) potassium chlorate (4) ammonium chlorate

6. Which kind of radiation will travel through an electric field on a pathway that remains unaffected by the field? (1) a proton (2) a gamma ray (3) an electron (4) an alpha particle

7. The major portion of an atom's mass consists of (1) electrons and protons (2) electrons and neutrons (3) neutrons and positrons (4) neutrons and protons

8. Which atom contains exactly 15 protons? (1) phosphorus-32 (2) sulfur-32 (3) oxygen-15 (4) nitrogen-15

9. Element X has two isotopes. If 72.0% of the element has an isotopic mass of 84.9 atomic mass units, and 28.0% of the element has an isotopic mass of 87.0 atomic mass units, the average atomic mass of element X is numerically equal to

    (1) $(72.0 + 84.9) \times (28.0 + 87.0)$

    (2) $(72.0 - 84.9) \times (28.0 + 87.0)$

    (3) $\dfrac{(72.0 \times 84.9)}{100} + \dfrac{(28.0 \times 87.0)}{100}$

    (4) $(72.0 \times 84.9) + (28.0 \times 87.0)$

10. Given the equation: $^{14}_{6}C \rightarrow ^{14}_{7}N + X$

    Which particle is represented by the letter X? (1) an alpha particle (2) a beta particle (3) a neutron (4) a proton

11. The atom of which element in the ground state has 2 unpaired electrons in the $2p$ sublevel? (1) fluorine (2) nitrogen (3) beryllium (4) carbon

12. Which atoms contain the same number of neutrons?

    (1) $^{1}_{1}H$ and $^{4}_{2}He$ (2) $^{2}_{1}H$ and $^{3}_{2}He$ (3) $^{3}_{1}H$ and $^{3}_{2}He$

    (4) $^{3}_{1}H$ and $^{4}_{2}He$

13. Which hydrocarbon formula is also an empirical formula?
(1) CH_4 (2) C_2H_4 (3) C_3H_6 (4) C_4H_8

14. The potential energy possessed by a molecule is dependent upon (1) its composition, only (2) its structure, only (3) both its composition and its structure (4) neither its composition nor its structure

15. Which is a correctly balanced equation for a reaction between hydrogen gas and oxygen gas?

    (1) $H_2(g) + O_2(g) \rightarrow H_2O(\ell) + heat$

    (2) $H_2(g) + O_2(g) \rightarrow 2H_2O(\ell) + heat$

    (3) $2H_2(g) + 2O_2(g) \rightarrow H_2O(\ell) + heat$

    (4) $2H_2(g) + O_2(g) \rightarrow 2H_2O(\ell) + heat$

16. The atom of which element has an ionic radius smaller than its atomic radius? (1) N   (2) S   (3) Br   (4) Rb

17. Which molecule contains a polar covalent bond?

(1) $\overset{\times\times}{\underset{\times\times}{\times}}\text{I}\overset{..}{\underset{..}{\times}}\text{I}\overset{..}{\underset{..}{}}:$ (3) $\text{H}\overset{..}{\underset{\overset{\times}{\bullet}}{\times}}\text{N}\overset{..}{}\text{H}$
H

(2) $\text{H}\overset{}{\times}\text{H}$ (4) $:\text{N}\overset{\times\times}{\underset{\times\times}{}}\text{N}\overset{\times}{\underset{\times}{}}$

18. Which is the correct formula for nitrogen (I) oxide? (1) NO (2) N_2O (3) NO_2 (4) N_2O_3

19. Which element in Group 15 has the strongest metallic character? (1) Bi (2) As (3) P (4) N

20. Which halogens are gases at STP?
(1) chlorine and fluorine (2) chlorine and bromine (3) iodine and fluorine (4) iodine and bromine

21. What is the total number of atoms represented in the formula $CuSO_4 \cdot 5H_2O$? (1) 8 (2) 13 (3) 21 (4) 27

22. When combining with nonmetallic atoms, metalic atoms generally will (1) lose electrons and form negative ions (2) lose electrons and form positive ions (3) gain electrons and form negative ions (4) gain electrons and form positive ions

23. Which set of elements contains a metalloid? (1) K, Mn, As, Ar (2) Li, Mg, Ca, Kr (3) Ba, Ag, Sn, Xe (4) Fr, F, O, Rn

24. Atoms of elements in a group on the Periodic Table have similar chemical properties. This similarity is most closely related to the atoms' (1) number of principal energy levels (2) number of valence electrons (3) atomic numbers (4) atomic masses

25. Which element in Period 2 of the Periodic Table is the most reactive nonmetal? (1) carbon (2) nitrogen (3) oxygen (4) fluorine

26. What is the gram formula mass of $(NH_4)_3PO_4$? (1) 113 g (2) 121 g (3) 149 g (4) 404 g

27. Given the reaction:

$CH_4 + 2O_2 \rightarrow CO_2 + 2H_2O$

What amount of oxygen is needed to completely react with 1 mole of CH_4? (1) 2 moles (2) 2 atoms (3) 2 grams (4) 2 molecules

28. Based on Reference Table E, which of the following saturated solutions would be the *least* concentrated? (1) sodium sulfate (2) potassium sulfate (3) copper (II) sulfate (4) barium sulfate

29. What is the total number of moles of H_2SO_4 needed to prepare 5.0 liters of a 2.0 M solution of H_2SO_4? (1) 2.5 (2) 5.0 (3) 10. (4) 20.

30. Given the reaction:

$Ca(s) + 2H_2O(\ell) \rightarrow Ca(OH)_2(aq) + H_2(g)$

When 40.1 grams of Ca(s) reacts completely with the water, what is the total volume, at STP, of $H_2(g)$ produced? (1) 1.00 L (2) 2.00 L (3) 22.4 L (4) 44.8 L

31. Which is the correct equilibrium expression for the reaction below?

$4NH_3(g) + 7O_2(g) \rightleftarrows 4NO_2(g) + 6H_2O(g)$

(1) $K = \dfrac{[NO_2][H_2O]}{[NH_3][O_2]}$ (3) $K = \dfrac{[NH_3][O_2]}{[NO_2][H_2O]}$

(2) $K = \dfrac{[NO_2]^4[H_2O]^6}{[NH_3]^4[O_2]^7}$ (4) $K = \dfrac{[NH_3]^4[O_2]^7}{[NO_2]^4[H_2O]^6}$

32. The potential energy diagram below shows the reaction $X + Y \rightleftarrows Z$.

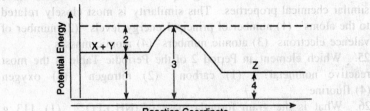

When a catalyst is added to the reaction, it will change the value of (1) 1 and 2 (2) 1 and 3 (3) 2 and 3 (4) 3 and 4

33. Which conditions will increase the rate of a chemical reaction? (1) decreased temperature and decreased concentration of reactants (2) decreased temperature and increased concentration of reactants (3) increased temperature and decreased concentration of reactants (4) increased temperature and increased concentration of reactants

34. A solution exhibiting equilibrium between the dissolved and undissolved solute must be (1) saturated (2) unsaturated (3) dilute (4) concentrated

35. Which 0.1 M solution has a pH greater than 7? (1) $C_6H_{12}O_6$ (2) CH_3COOH (3) KCl (2) KOH

36. What color is phenolphthalein in a basic solution? (1) blue (2) pink (3) yellow (4) colorless

37. According to Reference Table L, which of the following is the strongest Brönsted-Lowry acid? (1) HS^- (2) H_2S (3) HNO_2 (4) HNO_3

38. When HCl(aq) is exactly neutralized by NaOH(aq), the hydrogen ion concentration in the resulting mixture is (1) always less than the concentration of the hydroxide ions (2) always greater than the concentration of the hydroxide ions (3) always equal to the concentration of the hydroxide ions (4) sometimes greater and sometimes less than the concentration of the hydroxide ions

39. If 20. milliliters of 4.0 M NaOH is exactly neutralized by 20. milliliters of HCl, the molarity of the HCl is (1) 1.0 M (2) 2.0 M (3) 5.0 M (4) 4.0 M

40. The value of the ionization constant of water, K_w, will change when there is a change in (1) temperature (2) pressure (3) hydrogen ion concentration (4) hydroxide ion concentration

41. Based on Reference Table L, which species is amphoteric? (1) NH_2^- (2) NH_3 (3) I^- (4) HI

42. A redox reaction is a reaction in which (1) only reduction occurs (2) only oxidation occurs (3) reduction and oxidation occur at the same time (4) reduction occurs first and then oxidation occurs

43. Given the reaction:

$$\_Mg + \_Cr^{3+} \rightarrow \_Mg^{2+} + \_Cr$$

When the equation is correctly balanced using smallest whole numbers, the sum of the coefficients will be (1) 10 (2) 7 (3) 5 (4) 4

44. Oxygen has an oxidation number of -2 in (1) O_2 (2) NO_2 (3) Na_2O_2 (4) OF_2

45. Given the statements:

A The salt bridge prevents electrical contact between solutions of half-cells.

B The salt bridge prevents direct mixing of one half-cell solution with the other.

C The salt bridge allows electrons to migrate from one half-cell to the other.

D The salt bridge allows ions to migrate from one half-cell to the other.

Which two statements explain the purpose of a salt bridge used as part of a chemical cell? (1) A and C (2) A and D (3) C and D (4) B and D

46. When a substance is oxidized, it (1) loses protons (2) gains protons (3) acts as an oxidizing agent (4) acts as a reducing agent

47. In the reaction $Cu + 2Ag^+ \rightarrow Cu^2 + 2Ag$, the oxidizing agent is (1) Cu (2) Cu^{2+} (3) Ag^+ (4) Ag

48. A compound that is classified as organic must contain the element (1) carbon (2) nitrogen (3) oxygen (4) hydrogen

49. Which substance is a product of a fermentation reaction? (1) glucose (2) zymase (3) ethanol (4) water

50. Which of the following hydrocarbons has the *lowest* normal boiling point? (1) ethane (2) propane (3) butane (4) pentane

51. What type of reaction is $CH_3CH_3 + Cl_2 \rightarrow CH_3CH_2Cl + HCl$? (1) an addition reaction (2) a substitution reaction (3) a saponification reaction (4) an esterification reaction

52. Which compound is a saturated hydrocarbon? (1) ethane (2) ethene (3) ethyne (4) ethanol

Note that questions 53 through 56 have only three choices.

53. As atoms of elements in Group 16 are considered in order from top to bottom, the electronegativity of each successive element (1) decreases (2) increases (3) remains the same

54. As the pressure of a gas at 760 torr is changed to 380 torr at constant temperature, the volume of the gas (1) decreases (2) increases (3) remains the same

55. Given the change of phase: $CO_2(g) \rightarrow CO_2(s)$

As $CO_2(g)$ changes to $CO_2(s)$, the entropy of the system (1) decreases (2) increases (3) remains the same

56. In heterogeneous reactions, as the surface area of the reactants increases, the rate of the reaction (1) decreases (2) increases (3) remains the same

Part II

This part consists of twelve groups, each containing five questions. Each group tests a major area of the course. Choose seven of these twelve groups. Be sure that you answer all five questions in each group chosen. Record the answers to these questions on the separate answer sheet. [35]

Group I - Matter and Energy

If you choose this group, be sure to answer questions 57-61.

57. What is the total number of calories of heat energy absorbed by 15 grams of water when it is heated from 30.°C to 40.° C? (1) 10.   (2) 15 (3) 25 (4) 150

58. The graph below represents the uniform cooling of a sample of a substance, starting with the substance as a gas above its boiling point.

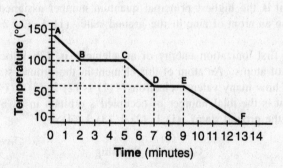

Which segment of the curve represents a time when both the liquid and the solid phases are present? (1) *EF*   (2) *BC*   (3) *CD* (4) *DE*

59. Which change of phase is exothermic?

   (1) $NaCl(s) \rightarrow NaCl(\ell)$ (2) $CO_2(s) \rightarrow CO_2(g)$

   (3) $H_2O(\ell) \rightarrow H_2O(s)$ (4) $H_2O(\ell) \rightarrow H_2O(g)$

60. According to the kinetic theory of gases, which assumption is correct? (1) Gas particles strongly attract each other. (2) Gas particles travel in curved paths. (3) The volume of gas particles prevents random motion. (4) Energy may be transferred between colliding particles.

61. A compound differs from a mixture in that a compound always has a (1) homogeneous composition (2) maximum of two components (3) minimum of three components (4) heterogeneous composition

Group 2- Atomic Structure

If you choose this group, be sure to answer questions 62-66.

62. An ion with 5 protons, 6 neutrons, and a charge of 3+ has an atomic number of (1) 5 (2) 6 (3) 8 (4) 11

63. Electron X can change to a higher energy level or a lower energy level. Which statement is true of electron X? (1) Electron X emits energy when it changes to a higher energy level. (2) Electron X absorbs energy when it changes to a higher energy level. (3) Electron X absorbs energy when it changes to a lower energy level. (4) Electron X neither emits nor absorbs energy when it changes energy level.

64. What is the highest principal quantum number assigned to an electron in an atom of zinc in the ground state (1) 1 (2) 2 (3) 5 (4) 4

65. The first ionization energy of an element is 176 kilocalories per mole of atoms. An atom of this element in the ground state has a total of how many valence electrons? (1) 1 (2) 2 (3) 3 (4) 4

66. What is the total number of occupied s orbitals in an atom of nickel in the ground state? (1) 1 (2) 2 (3) 3 (4) 4

Group 3 - Bonding

If you choose this group, be sure to answer questions 67-71.

67. What is the chemical formula for nickel (II) hypochlorite? (1) $NiCl_2$ (2) $Ni(ClO)_2$ (3) $NiClO_2$ (4) $Ni(ClO)_3$

68. Based on Reference Table G, which of the following compounds is most stable? (1) $CO(g)$ (2) $CO_2(g)$ (3) $NO(g)$ (4) $NO_2(g)$

69. The attractions that allow molecules of krypton to exist in the solid phase are due to (1) ionic bonds (2) covalent bonds (3) molecule-ion forces (4) van der Waals forces

70. Oxygen, nitrogen, and fluorine bond with hydrogen to form molecules. These molecules are attracted to each other by (1) ionic bonds (2) hydrogen bonds (3) electrovalent bonds (4) coordinate covalent bonds

71. An atom of which of the following elements has the greatest ability to attract electrons? (1) silicon (2) sulfur (3) nitrogen (4) bromine

Group 3 - Periodic Table

If you choose this group, be sure to answer questions 72-76.

72. Which electron configuration represents the atom with the largest covalent radius? (1) $1s^1$ (2) $1s^2 2s^1$ (3) $1s^2 2s^2$ (4) $1s^2 2s^2 2p^1$

73. A solution of $Cu(NO_3)_2$ is colored because of the presence of the ion (1) Cu^{2+} (2) N^{5+} (3) O^{2-} (4) NO_3^{1-}

74. Which element is more reactive than strontium? (1) potassium (2) calcium (3) iron (4) copper

75. At STP, which substance is the best conductor of electricity? (1) nitrogen (2) neon (3) sulfur (4) silver

76. The oxide of metal X has the formula XO. Which group in the Periodic Table contains metal X? (1) Group 1 (2) Group 2 (3) Group 13 (4) Group 17

Group 5 - Mathematics of Chemistry

If you choose this group, be sure to answer questions 77-81.

77. Given the same conditions of temperature and pressure, which noble gas will diffuse most rapidly? (1) He (2) Ne (3) Ar (4) Kr

78. What is the total number of molecules of hydrogen in 0.25 mole of hydrogen? (1) 6.0×10^{23} (2) 4.5×10^{23} (3) 3.0×10^{23} (4) 1.5×10^{23}

79. The volume of a 1.00-mole sample of an ideal gas will decrease when the (1) pressure decreases and the temperature decreases (2) pressure decreases and the temperature increases (3) pressure increases and the temperature decreases (4) pressure increases and the temperature increases

80. A 0.100-molal aqueous solution of which compound has the *lowest* freezing point? (1) $C_6H_{12}O_6$ (2) CH_3OH (3) $C_{12}H_{22}O_{11}$ (4) NaOH

81. What is the empirical formula of a compound that contains 85% Ag and 15% F by mass? (1) AgF (2) Ag_2F (3) AgF_2 (4) Ag_2F_2

82. Based on Reference Table *M*, which compound is less soluble in water than $PbCO_3$ at 298 K and 1 atmosphere? (1) AgI (2) AgCl (3) $CaSO_4$ (4) $BaSO_4$

83. Given the equilibrium reaction at constant pressure:

$$2HBr(g) + 17.4 \text{ kcal} \rightleftarrows H_2(g) + Br_2(g)$$

When the temperature is increased, the equilibrium will shift to the (1) right, and the concentration of HBr(g) will decrease (2) right, and the concentration of HBr(g) will increase (3) left, and the concentration of HBr(g) will decrease (4) left, and the concentration of HBr(g) will increase

84. A system is said to be in a state of dynamic equilibrium when the (1) concentration of products is greater than the concentration of reactants (2) concentration of products is the same as the concentration of reactants (3) rate at which products are formed is greater than the rate at which reactants are formed (4) rate at which products are formed is the same as the rate at which reactants are formed

85. Which reaction will occur spontaneously? [Refer to Reference Table G.]

(1) $\frac{1}{2}N_2(g) + \frac{1}{2}O_2(g) \rightarrow NO(g)$ (2) $\frac{1}{2}N_2(g) + O_2(g) \rightarrow NO_2(g)$

(3) $2C(s) + 3H_2(g) \rightarrow C_2H_6(g)$ (4) $2C(s) + 2H_2(g) \rightarrow C_2H_4(g)$

86. Which potential energy diagram represents the reaction $A + B \rightarrow C +$ energy?

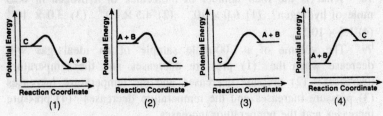

87. Potassium chloride, KCl, is a salt derived from the neutralization of a (1) weak acid and a weak base (2) weak acid and a strong base (3) strong acid and a weak base (4) strong acid and a strong base

88. Given the reaction

$$HSO_4^- + H_2O \rightleftarrows H_3O^+ + SO_4^{2-}$$

Which is a Brönsted-Lowry conjugate acid-base pair?
(1) HSO_4^- and H_3O^+ (2) HSO_4^- and SO_4^{2-} (3) H_2O and SO_4^{2-}
(4) H_2O and HSO_4^-

89. An aqueous solution that has a hydrogen ion concentration of 1.0×10^{-8} mole per liter has a pH of (1) 6, which is basic (2) 6, which is acidic (3) 8, which is basic (4) 8, which is acidic

90. The $[OH^-]$ of solution is 1×10^{-6}. At 298 K and 1 atmosphere, the product $[H_3O^+][OH^-]$ is (1) 1×10^{-2} (2) 1×10^{-6} (3) 1×10^{-8} (4) 1×10^{-14}

91. Given the reaction:
$$KOH + HNO_3 \rightarrow KNO_3 + H_2O$$
Which process is taking place?
(1) neutralization (2) esterification (3) substitution (4) addition

Group 8 - Redox and Electrochemistry

If you choose this group, be sure to answer questions 92-96.

92. Given the unbalanced equation:

$$\_MnO_2 + \_HCl \rightarrow \_MnCl_2 + \_H_2O + \_Cl_2$$

When the equation is correctly balanced using smallest whole-number coefficients, the coefficient of HCl is (1) 1 (2) 2 (3) 3 (4) 4

93. Based on Reference Table N, which half-cell has a lower electrode potential than the standard hydrogen half-cell?

(1) $Au^{3+} + 3e^- \rightarrow Au(s)$ (2) $Hg^{2+} + 2e^- \rightarrow Hg(\ell)$
(3) $Cu^+ + e^- \rightarrow Cu(s)$ (4) $Pb^{2+} + 2e^- \rightarrow Pb(s)$

94. According to Reference Table *N*, which reaction will take place spontaneously?
(1) $Ni^{2+} + Pb(s) \rightarrow Ni(s) + Pb^{2+}$ (2) $Au^{3+} + Al(s) \rightarrow Au(s) + Al^{3+}$
(3) $Sr^{2+} + Sn(s) \rightarrow Sr(s) + Sn^{2+}$ (4) $Fe^{2+} + Cu(s) \rightarrow Fe(s) + Cu^{2+}$

95. Given the reaction:

$$Mg(s) + Zn^{2+} \rightarrow (aq) \rightarrow Mg^{2+} \rightarrow (aq) + Zn(s)$$

What is the cell voltage (E^0) for the overall reaction?
(1) +1.61 V   (2) -1.61 V   (3) +3.13 V   (4) -3.13 V

96. The diagram below represents a chemical cell at 298 K.

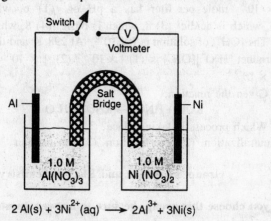

2 Al(s) + 3Ni^{2+}(aq) $\longrightarrow$ 2Al^{3+} + 3Ni(s)

When the switch is closed, electrons flow from
(1) Al(s) to Ni(s) (2) Ni(s) to Al(s) (3) Al^{3+}(aq) to Ni^{2+}(aq)
(4) Ni^{2+}(aq) to Al^{3+}(aq)

Group 9 - Organic Chemistry

If you choose this group, be sure to answer questions 97-101.

97. The compound C_4H_{10} belongs to the series of hydrocarbons with the general formula

(1) C_nH_{2n} (2) C_nH_{2n+2} (3) C_nH_{2n-2} (4) C_nH_{2n-6}

98. Which is an isomer of

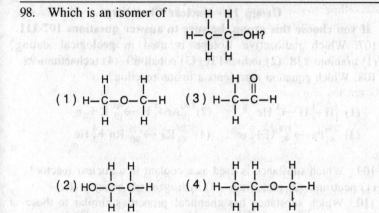

99. To be classified as a tertiary alcohol, the functional —OH group is bonded to a carbon atom that must be bonded to a total of how many additional carbon atoms? (1) 1 (2) 2 (3) 3 (4) 4

100. Which substance is made up of monomers joined together in long chains? (1) ketone (2) protein (3) ester (4) acid

101. What is the total number of carbon atoms in a molecule of glycerol? (1) 1 (2) 2 (3) 3 (4) 4

Group 10 - Applications of Chemical Principals
If you choose this group, be sure to answer questions 102-106.

102. Which type of reaction is occurring when a metal undergoes corrosion?
(1) oxidation-reduction (2) neutralization (3) polymerization (4) saponification

103. Which process is used to separate the components of a petroleum mixture?
(1) addition polymerization (2) condensation polymerization (3) fractional distillation (4) fractional crystallization

104. Which substance functions as the electrolyte in an automobile battery? (1) PbO_2 (2) $PbSO_4$ (3) H_2SO_4 (4) H_2O

105. A battery consists of which type of cells? (1) electrolytic (2) electrochemical (3) electroplating (4) electromagnetic

106. Which element can be found in nature in the free (uncombined) state? (1) Ca (2) Ba (3) Au (4) Al

Group 11 - Nuclear Chemistry
If you choose this group, be sure to answer questions 107-111.

107. Which radioactive isotope is used in geological dating?
(1) uranium-238 (2) iodine-131 (3) cobalt-60 (4) technetium-99

108. Which equation represents a fusion reaction?

(1) $_1^3H + _1^1H \rightarrow _2^4He$ (2) $_{18}^{40}Ar + _1^1H \rightarrow _{19}^{40}K + _0^1n$

(3) $_{91}^{234}Pa \rightarrow _{92}^{234}U + _1^0e$ (4) $_{88}^{226}Ra \rightarrow _{86}^{226}Rn + _2^4He$

109. Which substance is used as a coolant in a nuclear reactor?
(1) neutrons (2) plutonium (3) hydrogen (4) heavy water

110. Which substance has chemical properties similar to those of radioactive ^{235}U? (1) ^{235}Pa (2) ^{233}Pa (3) ^{233}U (4) ^{206}Pb

111. Control rods in nuclear reactors are commonly made of boron and cadmium because these two elements have the ability to
(1) absorb neutrons (2) emit neutrons (3) decrease the speed of neutrons (4) increase the speed of neutrons

Group 12 - Laboratory Activities
If you choose this group, be sure to answer questions 112-116.

Base your answers to questions 112 and 113 on the table below, which represents the production of 50 milliliters of CO_2 in the reaction of HCl with $NaHCO_3$. Five trials were performed under different conditions as shown. (The same mass of $NaHCO_3$ was used in each trial.

Trial	Particle Size of $NaHCO_3$	Concentration of HCl	Temperature (°C) of HCl
A	small	1 M	20
B	large	1 M	20
C	large	1 M	40
D	small	2 M	40
E	large	2 M	40

112. Which two trials could be used to measure the effect of surface area? (1) trials A and B (2) trials A and C (3) trials A and D (4) trials B and D

113. Which trial would produce the fastest reaction? (1) trial A (2) trial B (3) trial C (4) trial D

114. A student determined the heat of fusion of water to be 88 calories per gram. If the accepted value is 80. calories per gram, what is the student's percent error? (1) 8.0% (2) 10.% (3) 11% (4) 90.%

115. Given: (52.6 cm) (1.214 cm)

What is the product expressed to the correct number of significant figures? (1) 64 cm^2 (2) 63.9 cm^2 (3) 63.86 cm^2 (4) 63.8564 cm^2

116. The diagram below represents a metal bar and two centimeter rulers, A and B. Portions of the rulers have been enlarged to show detail.

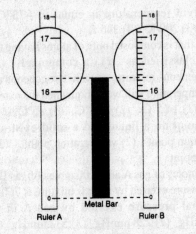

Ruler A   Metal Bar   Ruler B

What is the greatest degree of precision to which the metal bar can be measured by ruler A and by ruler B? (1) to the nearest tenth by both rulers (2) to the nearest hundredth by both rulers (3) to the nearest tenth by ruler A and to the nearest hundredth by ruler B (4) to the nearest hundredth by ruler A and to the nearest tenth by ruler B

CHEMISTRY
June 20, 1997

Part I
Answer all 56 questions in this part. [65]

Directions (1-56): For *each* statement or question, select the word or expression that, of those given, best completes the statement or answers the question. Record your answer on the separate answer sheet in accordance with the directions on the front page of this booklet.

1. Which Kelvin temperature is equal to $-73°C$? (1) 100 K (2) 173 K (3) 200 K (4) 346 K

2. A substance that is composed only of atoms having the same atomic number is classified as (1) a compound (2) an element (3) a homogeneous mixture (4) a heterogeneous mixture

3. At which temperature will water boil when the external pressure is 17.5 torr? (1) 14.5°C (2) 16.5°C (3) 20°C (4) 100°C

4. At which point do a liquid and a solid exist at equilibrium? (1) sublimation point (2) vaporization point (3) boiling point (4) melting point

5. When 7.00 moles of gas A and 3.00 moles of gas B are combined, the total pressure exerted by the gas mixture is 760. mmHg. What is the partial pressure exerted by gas A in this mixture? (1) 76.0 mmHg (2) 228 mmHg (3) 532 mmHg (4) 760. mmHg

6. Which radioactive emanations have a charge of 2+? (1) alpha particles (2) beta particles (3) gamma rays (4) neutrons

7. Which symbols represent atoms that are isotopes of each other? (1) ^{14}C and ^{14}N (2) ^{16}O and ^{18}O (3) ^{131}I and ^{131}I (4) ^{222}Rn and ^{222}Ra

8. Which orbital notation correctly represents the outermost principal energy level of a nitrogen atom in the ground state?

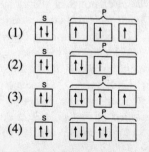

9. The atomic mass of an element is defined as the weighted average mass of that element's (1) most abundant isotope (2) least abundant isotope (3) naturally occurring isotopes (4) radioactive isotopes

10. When electrons in an atom in an excited state fall to lower energy levels, energy is (1) absorbed, only (2) released, only (3) neither released nor absorbed (4) both released and absorbed

11. A neutron has approximately the same mass as (1) an alpha particle (2) a beta particle (3) an electron (4) a proton

12. What is the formula for sodium oxalate? (1) $NaClO$ (2) Na_2O (3) $Na_2C_2O_4$ (4) $NaC_2H_3O_2$

13. Given the unbalanced equation: $Al + O_2 \rightarrow Al_2O_3$
When this equation is completely balanced using smallest whole numbers, what is the sum of the coefficient?
(1) 9 (2) 7 (3) 5 (4) 4

14. One mole of which substance contains a total of 6.02×10^{23} atoms? (1) Li (2) NH_3 (3) O_2 (4) CO_2

15. Which formula represents a molecular substance?
(1) CaO (2) CO (3) Li_2O (4) Al_2O_3

16. In an aqueous solution of an ionic salt, the oxygen atom of the water molecule is attracted to the
(1) negative ion of the salt, due to oxygen's partial positive charge
(2) negative ion of the salt, due to oxygen's partial negative charge
(3) positive ion of the salt, due to oxygen's partial positive charge
(4) positive ion of the salt, due to oxygen's partial negative charge

17. What is the empirical formula of the compound whose molecular formula is P_4O_{10}? (1) PO (2) PO_2 (3) P_2O_5 (4) P_8O_{20}

18. Which sequence of Group 18 elements demonstrates a gradual *decrease* in the strength of the van der Waals forces?

(1) $Ar(\ell), Kr(\ell), Ne(\ell), Xe(\ell)$ (2) $Kr(\ell), Xe(\ell), Ar(\ell), Ne(\ell)$

(3) $Ne(\ell), Ar(\ell), Kr(\ell), Xe(\ell)$ (4) $Xe(\ell), Kr(\ell), Ar(\ell), Ne(\ell)$

19. In the ground state, atoms of the elements in Group 15 of the Periodic Table all have the same number of (1) filled principal energy levels (2) occupied principal energy levels (3) neutrons in the nucleus (4) electrons in the valence shell

20. Which elements have the most similar chemical properties?
(1) K and Na (2) K and Cl (3) K and Ca (4) K and S

21. Which three groups of the Periodic Table contain the most elements classified as metalloids (semimetals)? (1) 1, 2, and 13 (2) 2, 13, and 14 (3) 14, 15, and 16 (4) 16, 17, and 18

22. In which classification is an element placed if the outermost 3 sublevels of its atoms have a ground state electron configuration of $3p^63d^54s^2$? (1) alkaline earth metals (2) transition metals (3) metalloids (semimetals) (4) nonmetals

23. A diatomic element with a high first ionization energy would most likely be a (1) nonmetal with a high electronegativity (2) nonmetal with a low electronegativity (3) metal with a high electronegativity (4) metal with a low electronegativity

24. As the elements in Period 3 are considered from left to right, they tend to (1) lose electrons more readily and increase in metallic character (2) lose electrons more readily and increase in nonmetallic character (3) gain electrons more readily and increase in metallic character (4) gain electrons more readily and increase in nonmetallic character

25. An atom of an element has 28 innermost electrons and 7 outermost electrons. In which period of the Periodic Table is this element located? (1) 5 (2) 2 (3) 3 (4) 4

26. Which solution is the most concentrated? (1) 1 mole of solute dissolved in 1 liter of solution (2) 2 moles of solute dissolved in 3 liters of solution (3) 6 moles of solute dissolved in 4 liters of solution (4) 4 moles of solute dissolved in 8 liters of solution

27. What is the gram formula mass of K_2CO_3?
    (1) 138 g (2) 106 g (3) 99 g (4) 67 g

28. What is the total number of atoms contained in 2.00 moles of nickel?
    (1) 58.9 (2) 118 (3) 6.02×10^{23} (4) 1.20×10^{24}

29. Given the reaction at STP: $2KClO_3(s) \rightarrow 2KCl(s) + 3O_2(g)$
    What is the total number of liters of $O_2(g)$ produced from the complete decomposition of 0.500 mole of $KClO_3(s)$?
    (1) 11.2 L (2) 16.8 L (3) 44.8 L (4) 67.2 L

30. What is the percent by mass of oxygen in magnesium oxide, MgO?
    (1) 20% (2) 40% (3) 50% (4) 60%

31. A solution in which the crystallizing rate of the solute equals the dissolving rate of the solute must be (1) saturated (2) unsaturated (3) concentrated (4) dilute

32. Which statement explains why the speed of some chemical reactions is increased when the surface area of the reactant is increased? (1) This change increases the density of the reactant particles. (2) This change increases the concentration of the reactant. (3) This change exposes more reactant particles to a possible collision. (4) This change alters the electrical conductivity of the reactant particles.

33. According to Reference Table G, which compound forms exothermically? (1) hydrogen fluoride (2) hydrogen iodide (3) ethene (4) ethyne

34. The potential energy diagram shown below represents the reaction $A + B \rightarrow AB$.

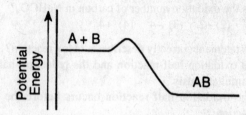

Which statement correctly describes this reaction?
(1) It is endothermic and energy is absorbed.
(2) It is endothermic and energy is released.
(3) It is exothermic and energy is absorbed.
(4) It is exothermic and energy is released.

35. Given the reaction at equilibrium: $N_2(g) + 3H_2(g) \rightleftarrows 2NH_3(g)$

Increasing the concentration of $N_2(g)$ will increase the forward reaction rate due to (1) a decrease in the number of effective collisions (2) an increase in the number of effective collisions (3) a decrease in the activation energy (4) an increase in the activation energy

36. Based on Reference Table L, which of the following aqueous solutions is the best conductor of electricity?
(1) 0.1 M HF (2) 0.1 M H_2S (3) 0.1 M H_2SO_4 (4) 0.1 M H_3PO_4

37. Which substance is classified as an Arrhenius base?
(1) HCl (2) NaOH (3) $LiNO_3$ (4) $KHCO_3$

38. The conjugate acid of the HS^- ion is (1) H^+ (2) S^{2-} (3) H_2O (4) H_2S

39. If 20. milliliters of 1.0 M HCl was used to completely neutralize 40. milliliters of an NaOH solution, what was the molarity of the NaOH solution? (1) 0.50 M (2) 2.0 M (3) 1.5 M (4) 4.0 M

40. According to Reference Table L, which species is amphoteric (amphiprotic)? (1) HCl (2) HNO_2 (3) HSO_4^- (4) H_2SO_4

41. In the reaction $H_2O + CO_3^{2-} \rightleftarrows OH^- + HCO_3^-$, the two Brönsted-Lowry acids are (1) H_2O and OH^- (2) H_2O and HCO_3^- (3) CO_3^{2-} and OH^- (4) CO_3^{2-} and HCO_3^-

42. What happens to reducing agents in chemical reactions?
(1) Reducing agents gain protons. (2) Reducing agents gain electrons (3) Reducing agents are oxidized. (4) Reducing agents are reduced

43. What is the oxidation number of carbon in $NaHCO_3$?
(1) +6 (2) +2 (3) –4 (4) +4

44. Which statement correctly describes a redox reaction?
(1) The oxidation half-reaction and the reduction half-reaction occur simultaneously.
(2) The oxidation half-reaction occurs before the reduction half-reaction.
(3) The oxidation half-reaction occurs after the reduction half-reaction.
(4) The oxidation half-reaction occurs spontaneously but the reduction half-reaction does not.

45. Given the redox reaction: $Co(s) + PbCl_2(aq) \rightarrow CoCl_2(aq) + Pb(s)$
Which statement correctly describes the oxidation and reduction that occur?

(1) $Co(s)$ is oxidized and $Cl^-(aq)$ is reduced.
(2) $Co(s)$ is oxidized and $Pb^{2+}(aq)$ is reduced.
(3) $Co(s)$ is reduced and $Cl^-(aq)$ is oxidized.
(4) $Co(s)$ is reduced and $Pb^{2+}(aq)$ is oxidized.

46. Which half-reaction correctly represents reduction?

(1) $Cr^{3+} + 3e^- \rightarrow Cr(s)$ (2) $Cr^{3+} \rightarrow Cr(s) + 3e^-$

(3) $Cr(s) \rightarrow Cr^{3+} + 3e^-$ (4) $Cr(s) + 3e^- \rightarrow Cr^{3+}$

47. Which statement best describes how a salt bridge maintains electrical neutrality in the half-cells of an electrochemical cell?

(1) It prevents the migration of electrons.
(2) It permits the migration of ions.
(3) It permits the two solutions to mix completely.
(4) It prevents the reaction from occurring spontaneously.

48. What is the name of a compound that has the molecular formula C_6H_6? (1) butane (2) butene (3) benzene (4) butyne

49. The fermentation of $C_6H_{12}O_6$ will produce CO_2 and
(1) $C_3H_5(OH)_3$ (2) C_2H_5OH (3) $Ca(OH)_2$ (4) $Cr(OH)_3$

50. Which is the correct name for the substance below?

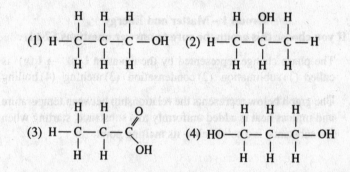

   (1) ethanol   (2) ethyne   (3) ethane   (4) ethene

51. Which structural formula represents an organic acid?

52. In a molecule of CH_4, the hydrogen atoms are spatially oriented toward the corners of a regular (1) pyramid (2) tetrahedron (3) square (3) rectangle

Note that questions 53 through 56 have only three choices.

53. Given the reaction at equilibrium: $N_2(g) + O_2(g) \rightleftarrows 2NO(g)$
As the concentration of $N_2(g)$ increases, the concentration of $O_2(g)$ will (1) decrease (2) increase (3) remain the same

54. As the temperature of a sample of a radioactive element decreases, the half-life of the element will

   (1) decrease   (2) increase   (3) remain the same

55. As ice cools from 273 K to 263 K, the average kinetic energy of its molecules will

   (1) decrease   (2) increase   (3) remain the same

56. As the hydrogen ion concentration of an aqueous solution increases, the hydroxide ion concentration of this solution will

   (1) decrease   (2) increase   (3) remain the same

Part II

This part consists of twelve groups, each containing five questions. Each group tests a major area of the course. Choose seven of these twelve groups. Be sure that you answer all five questions in each group chosen. Record the answers to these questions on the separate answer sheet in accordance with the directions on the front page of this booklet. [35]

Group 1 – Matter and Energy
If you choose this group, be sure to answer questions 57-61.

57. The phase change represented by the equation $I_2(s) \rightarrow I_2(g)$ is called (1) sublimation (2) condensation (3) melting (4) boiling

58. The graph below represents the relationship between temperature and time as heat is added uniformly to a substance, starting when the substance is a solid below its melting point.

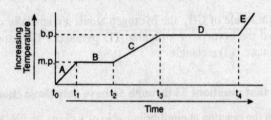

Which portions of the graph represent times when heat is absorbed and potential energy increases while kinetic energy remains constant? (1) A and B (2) B and D (3) A and C (4) C and D

59. The heat of fusion is defined as the energy required at constant temperature to change 1 unit mass of a (1) gas to a liquid (2) gas to a solid (3) solid to a gas (4) solid to a liquid

60. Given the equation: $2Na + 2H_2O \rightarrow 2NaOH + H_2$
Which substance in this equation is a binary compound?
(1) Na (2) H_2 (3) H_2O (4) NaOH

61. At STP, 1 liter of $O_2(g)$ and 1 liter of $Ne(g)$ have the same (1) mass (2) density (3) number of atoms (4) number of molecules

Group 2 – Atomic Structure
If you choose this group, be sure to answer questions 62-66.

62. The diagram below represents radiation passing through an electric field.

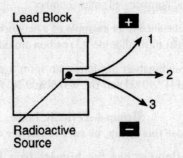

Which type of emanation is represented by the arrow labeled 2?
(1) alpha particle   (2) beta particle   (3) positron   (4) gamma radiation

63. Which sample will decay *least* over a period of 30 days? [Refer to Reference Table H.] (1) 10 g of Au-198 (2) 10 g of I-131 (3) 10 g of P-32 (4) 10 g of Rn-222

64. A particle has a mass of 1.0 atomic mass unit. What is the approximate mass of this particle in grams? (1) 1.0 g (2) 2.0 g (3) 1.7×10^{-24} g   (4) 6.0×10^{-23} g

65. Which equation represents nuclear disintegration resulting in release of beta particle?

(1) $^{220}_{87}Fr + ^{4}_{2}He \rightarrow ^{224}_{89}Ac$ (2) $^{239}_{94}Pu \rightarrow ^{235}_{92}U + ^{4}_{2}He$

(3) $^{32}_{15}P + ^{0}_{-1}e \rightarrow ^{32}_{14}Si$ (4) $^{198}_{79}Au \rightarrow ^{198}_{80}Hg + ^{0}_{-1}e$

66. Which electron configuration represents a potassium atom in the excited state?
(1) $1s^2 2s^2 2p^6 3s^2 3p^3$ (2) $1s^2 2s^2 2p^6 3s^1 3p^4$
(3) $1s^2 2s^2 2p^6 3s^2 3p^6 4s^1$ (4) $1s^2 2s^2 2p^6 3s^2 3p^5 4s^2$

Group 3 – Bonding
If you choose this group, be sure to answer questions 67-71.

67. Which type of attraction is directly involved when KCl dissolves in water?   (1) molecule–molecule   (2) molecule–atom (3) molecule–ion   (4) ion–ion

68. In which compound have electrons been transferred to the oxygen atom? (1) CO_2 (2) NO_2 (3) N_2O (4) Na_2O

69. A strontium atom differs from a strontium ion in that the atom has a greater (1) number of electrons (2) number of protons (3) atomic number (4) mass number

70. Which substance is an example of a network solid? (1) nitrogen dioxide (2) sulfur dioxide (3) carbon dioxide (4) silicon dioxide

71. Which combination of atoms can form a polar covalent bond? (1) H and H (2) H and Br (3) N and N (4) Na and Br

Group 4–Periodic Table
If you choose this group, be sure to answer questions 72-76.

72. Which element has the highest first ionization energy? (1) sodium (2) aluminum (3) calcium (4) phosphorus

73. Which compound forms a colored aqueous solution? (1) $CaCl_2$ (2) $CrCl_3$ (3) NaOH (4) KBr

74. When a metal atom combines with a nonmetal atom, the nonmetal atom will (1) lose electrons and decrease in size (2) lose electrons and increase in size (3) gain electrons and decrease in size (4) gain electrons and increase in size

75. According to Reference Table P, which of the following elements has the smallest covalent radius? (1) nickel (2) cobalt (3) calcium (4) potassium

76. Which element's ionic radius is smaller than its atomic radius? (1) neon (2) nitrogen (3) sodium (4) sulfur

Group 5–Mathematics of Chemistry
If you choose this group, be sure to answer questions 77-81.

77. What is the total number of moles of hydrogen gas contained in 9.03×10^{23} molecules? (1) 1.50 moles (2) 2.00 moles (3) 6.02 moles (4) 9.03 moles

78. At the same temperature and pressure, which gas will diffuse through air at the fastest rate? (1) H_2 (2) O_2 (3) CO (4) CO_2

79. How are the boiling and freezing points of a sample of water affected when a salt is dissolved in the water?
   (1) The boiling point decreases and the freezing point decreases.
   (2) The boiling point decreases and the freezing point increases.
   (3) The boiling point increases and the freezing point decreases.
   (4) The boiling point increases and the freezing point increases.

80. A sample of an unknown gas at STP has a density of 0.630 gram per liter. What is the gram molecular mass of this gas? (1) 2.81 g (2) 14.1 g (3) 22.4 g (4) 63.0 g

81. A compound is 86% carbon and 14% hydrogen by mass. What is the empirical formula for this compound? (1) CH (2) CH_2 (3) CH_3 (4) CH_4

Group 6–Kinetics and Equilibrium

If you choose this group, be sure to answer questions 82-86.

82. In a chemical reaction, a catalyst changes the (1) potential energy of the products (2) potential energy of the reactants (3) heat of reaction (4) activation energy

83. Which statement describes characteristics of an endothermic reaction? (1) The sign of ΔH is positive, and the products have less potential energy than the reactants. (2) The sign of ΔH is positive, and the products have more potential energy than the reactants. (3) The sign of ΔH is negative, and the products have less potential energy than the reactants. (4) The sign of ΔH is negative, and the products have more potential energy than the reactants.

84. What is the K_{sp} expression for the salt PbI_2?
(1) $[Pb^{2+}][I^-]^2$ (2) $[Pb^{2+}][2I^-]$ (3) $[Pb^{2+}][I_2]^2$ (4) $[Pb^{2+}][2I^-]^2$

85. Given the equilibrium system: $PbCO_3(s) \rightleftarrows Pb^{2+}(aq) + CO_3^{2-}(aq)$

Which changes occur as $Pb(NO_3)_2(s)$ is added to the system at equilibrium?
(1) The amount of $PbCO_3(s)$ decreases, and the concentration of $CO_3^{2-}(aq)$ decreases.
(2) The amount of $PbCO_3(s)$ decreases, and the concentration of $CO_3^{2-}(aq)$ increases.
(3) The amount of $PbCO_3(s)$ increases, and the concentration of $CO_3^{2-}(aq)$ decreases.
(4) The amount of $PbCO_3(s)$ increases, and the concentration of $CO_3^{2-}(aq)$ increases.

86. A chemical reaction will always occur spontaneously if the reaction has a negative (1) ΔG (2) ΔH (3) ΔS (4) T

Group 7–Acids and Bases

If you choose this group, be sure to answer questions 87-91.

87. An acidic solution could have a pH of (1) 7 (2) 10 (3) 3 (4) 14

88. What is the pH of a 0.00001 molar HCl solution?
(1) 1 (2) 9 (3) 5 (4) 4

89. According to the Brönsted-Lowry theory, an acid is any species that can (1) donate a proton (2) donate an electron (3) accept a proton (4) accept an electron

90. When the salt Na_2CO_3 undergoes hydrolysis, the resulting solution will be (1) acidic with a pH less than 7 (2) acidic with a pH greater than 7 (3) basic with a pH less than 7 (4) basic with a pH greater than 7

91. In an aqueous solution, which substance yields hydrogen ions as the only positive ions? (1) C_2H_5OH (2) CH_3COOH (3) KH (4) KOH

Group 8–Redox and Electrochemistry

If you choose this group, be sure to answer questions 92-96.

92. In which kind of cell are the redox reactions made to occur by an externally applied electrical current? (1) galvanic cell (2) chemical cell (3) electrochemical cell (4) electrolytic cell

93. According to Reference Table N, which metal will react spontaneously with Ag^+ ions, but not with Zn^{2+} ions?
(1) Cu (2) Au (3) Al (4) Mg

94. Which atom forms an ion that would migrate toward the cathode in an electrolytic cell? (1) F (2) I (3) Na (4) Cl

95. Given the equations A, B, C, and D:

(A) $AgNO_3 + NaCl \rightarrow AgCl + NaNO_3$
(B) $Cl_2 + H_2O \rightarrow HClO + HCl$
(C) $CuO + CO \rightarrow CO_2 + Cu$
(D) $NaOH + HCl \rightarrow NaCl + H_2O$

Which two equations represent redox reactions?

(1) A and B (2) B and C (3) C and A (4) D and B

96. Given the unbalanced equation:

$$\_NO_3^- + 4H^+ + \_Pb \rightarrow \_Pb^{2+} + \_NO_2 + 2H_2O$$

What is the coefficient of NO_2 when the equation is correctly balanced? (1) 1 (2) 2 (3) 3 (4) 4

Group 9–Organic Chemistry

If you choose this group, be sure to answer questions 97-101.

97. Which polymers occur naturally? (1) starch and nylon (2) starch and cellulose (3) protein and nylon (4) protein and plastic

98. Which statement explains why the element carbon forms so many compounds?
    (1) Carbon atoms combine readily with oxygen.
    (2) Carbon atoms have very high electronegativity.
    (3) Carbon readily forms ionic bonds with other carbon atoms.
    (4) Carbon readily forms covalent bonds with other carbon atoms.

99. Which structural formula represents a primary alcohol?

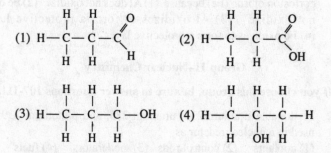

100. Which compounds are isomers? (1) 1-propanol and 2-propanol (2) methanoic acid and ethanoic acid (3) methanol and methanal (4) ethane and ethanol

101. Compared to the rate of inorganic reactions, the rate of organic reactions generally is
    (1) slower because organic particles are ions
    (2) slower because organic particles contain covalent bonds
    (3) faster because organic particles are ions
    (4) faster because organic particles contain covalent bonds

Group 10–Applications of Chemical Principals

If you choose this group, be sure to answer questions 102-106.

102. Which products are obtained from the fractional distillation of petroleum? (1) esters and acids (2) alcohols and aldehydes (3) soaps and starches (4) kerosene and gasoline

103. Given the lead-acid battery reaction:

$$Pb + PbO_2 + 2H_2SO_4 \underset{charge}{\overset{discharge}{\rightleftharpoons}} 2PbSO_4 + 2H_2O$$

Which species is oxidized during battery discharge?
(1) Pb   (2) PbO_2 (3) SO_4^{2-} (4) H_2O

104. Given the reaction: $ZnO + X + heat \rightarrow Zn + XO$
Which element, represented by X, is used industrially to reduce the ZnO to Zn? (1) Cu   (2) C   (3) Sn   (4) Pb

105. Which metal is obtained commercially by the electrolysis of its salt? (1) Zn   (2) K   (3) Fe   (4) Ag

106. The corrosion of aluminum (Al) is a less serious problem than the corrosion of iron (Fe) because (1) Al does not oxidize (2) Fe does not oxidize   (3) Al oxidizes to form a protective layer (4) Fe oxidizes to form a protective layer

Group 11–Nuclear Chemistry

If you choose this group, be sure to answer questions 107-111.

107. Fissionable uranium-233, uranium-235, and plutonium-239 are used in a nuclear reactor as
(1) coolants   (2) control rods   (3) moderators   (4) fuels

108. Which reaction illustrates fusion?

(1) $_1^2H + _1^2H \rightarrow _2^4He$

(2) $_0^1n + _{13}^{27}Al \rightarrow _{11}^{24}Na + _2^4He$

(3) $_{13}^{27}Al + _2^4He \rightarrow _{15}^{30}P + _0^1n$

(4) $_7^{14}N + _2^4He \rightarrow _1^1H + _8^{17}O$

109. An accelerator can *not* be used to speed up (1) alpha particles (2) beta particles (3) protons (4) neutrons

110. Brain tumors can be located by using an isotope of (1) carbon-14 (2) iondine-131 (3) technetium-99 (4) uranium-238

111. In the reaction $^{9}_{4}Be + X \rightarrow {}^{12}_{6}C + {}^{1}_{0}n$, the X represents
(1) an alpha particle   (2) a beta particle   (3) an electron (4) a proton

Group 12–Laboratory Activities
If you choose this group, be sure to answer questions 112-116

112. Which piece of laboratory equipment should be used to remove a heated crucible from a ring stand?

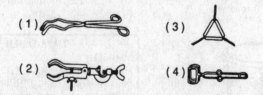

113. The following set of procedures was used by a student to determine the heat of solution of NaOH.
    (*A*) Read the original temperature of the water.
    (*B*) Read the final temperature of the solution.
    (*C*) Pour the water into a beaker.
    (*D*) Stir the mixture.
    (*E*) Add the sodium hydroxide.
What is the correct order of procedures for making this determination?

(1) A →C →E →B →D   (2) E →D →C →A →B

(3) C →A →E →D →B   (4) C →E →D →A →B

114. In an experiment, a student found 18.6% by mass of water in a sample of $BaCl_2 \cdot 2H_2O$. The accepted value is 14.8%. What was the student's experimental percent error?

(1) $\dfrac{3.8}{18.6} \times 100$ (2) $\dfrac{3.8}{14.8} \times 100$

(3) $\dfrac{14.8}{18.6} \times 100$ (4) $\dfrac{18.6}{14.8} \times 100$

115. A student obtained the following data in a chemistry laboratory.

Trial	Temperature (°C)	Solubility (grams of KNO_3/100 g of H_2O)
1	25	40
2	32	50
3	43	70
4	48	60

Based on Reference Table D, which of the trials seems to be in error? (1) 1 (2) 2 (3) 3 (4) 4

116. A student using a Styrofoam cup as a calorimeter added a piece of metal to distilled water and stirred the mixture as shown in the diagram below. The student's data is shown in the table below.

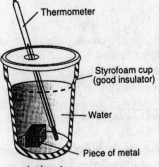

Thermometer

Styrofoam cup (good insulator)

Water

Piece of metal

calorimeter

DATA TABLE	
Mass of H_2O	50.0 g
Initial temperature of H_2O	25.0°C
Mass of metal	20.0 g
Initial temperature of metal	100.°C
Final temperature of H_2O + metal	32.0°C

Which statement correctly describes the heat flow in calories? [Ignore heat gained or lost by the calorimeter.]

(1) The water lost 1360 calories of heat and the metal gained 140. calories of heat.

(2) The water lost 350. calories of heat and the metal gained 350. calories of heat.

(3) The water gained 1360 calories of heat and the metal lost 140. calories of heat.

(4) The water gained 350. calories of heat and the metal lost 350. calories of heat.

CHEMISTRY
June 23, 1998

Part I
Answer all 56 questions in this part. [65]

Directions (1-56): For *each* statement or question, select the word or expression that, of those given, best completes the statement or answers the question. Record your answer on the separate answer sheet in accordance with the directions on the front page of this booklet.

1. The diagrams below represent two solids and the temperature of each.

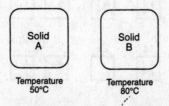

   What occurs when the two solids are placed in contact with each other? (1) Heat energy flows from solid *A* to solid *B*. Solid *A* decreases in temperature. (2) Heat energy flows from solid *A* to solid *B*. Solid *A* increases in temperature. (3) Heat energy flows from solid *B* to solid *A*. Solid *B* decreases in temperature. (4) Heat energy flows from solid *B* to solid *A*. Solid *B* increases in temperature.

2. The particles of a substance are arranged in a definite geometric pattern and are constantly vibrating. This substance can be in (1) the solid phase, only (2) the liquid phase, only (3) either the liquid or the solid phase (4) neither the liquid nor the solid phase

3. What is the pressure of a mixture of CO_2, SO_2, and H_2O gases, if each gas has a partial pressure of 250 torr?
   (1) 250 torr (2) 500 torr (3) 750 torr (4) 1000 torr

4. Which substances can be decomposed chemically?
   (1) CaO and Ca   (2) MgO and Mg
   (3) CO and Co   (4) CaO and MgO

5. A gas sample has a volume of 25.0 milliliters at a pressure of 1.00 atmosphere. If the volume increases to 50.0 milliliters and the temperature remains constant, the new pressure will be
   (1) 1.00 atm (2) 2.00 atm (3) 0.250 atm (4) 0.500 atm

6. At atom with the electron configuration $1s^2 2s^2 2p^6 3s^2 3p^6 3d^5 4s^2$ has an incomplete (1) $2p$ sublevel (2) second principal energy level (3) third principal energy level (4) $4s$ sublevel

7. Which orbital notation represents a boron atom in the ground state?

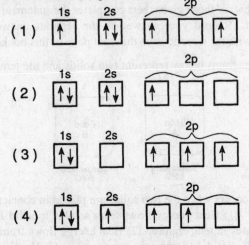

8. In the equation, $^{234}_{90}\text{Th} \rightarrow \,^{234}_{91}\text{Pa} + X$, the symbol X represents

(1) $^{0}_{+1}\text{e}$ (2) $^{0}_{-1}\text{e}$ (3) $^{1}_{0}\text{n}$ (4) $^{1}_{1}\text{H}$

9. Which subatomic particle is found in the nucleus of all isotopes of hydrogen?
(1) proton   (2) neutron   (3) electron   (4) positron

10. What is the highest principal quantum number (n) for an electron in an atom of sulfur in the ground state?
(1) 1   (2) 2   (3) 3   (4) 4

11. What is the total number of electrons in a completely filled fourth principal energy level?
(1) 8   (2) 10   (3) 16   (4) 32

12. What is the total number of hydrogen atoms required to form 1 molecule of $C_3H_5(OH)_3$?
(1) 1   (2) 5   (3) 3   (4) 8

13. Which element is found in both potassium chlorate and zinc nitrate?
(1) hydrogen   (2) oxygen   (3) potassium   (4) zinc

14. Which formula represents lead (II) phosphate?
(1) $PbPO_4$ (2) Pb_4PO_4 (3) $Pb_3(PO_4)_2$ (4) $Pb_2(PO_4)_3$

15. Atoms of which element have the *weakest* attraction for electrons?
(1) Na (2) P (3) Si (4) S

16. The ability to conduct electricity in the solid state is a characteristic of metallic bonding. This characteristic is best explained by the presence of
(1) high ionization energies (2) high electronegativities
(3) mobile electrons (4) mobile protons

17. When ionic bonds are formed, metallic atoms tend to
(1) lose electrons and become negative ions
(2) lose electrons and become positive ions
(3) gain electrons and become negative ions
(4) gain electrons and become positive ions

18. The bond between hydrogen and oxygen in a water molecule is classified as
(1) ionic and nonpolar (2) ionic and polar
(3) covalent and nonpolar (4) covalent and polar

19. According to the Periodic Table, which element has more than one positive oxidation state?
(1) cadmium (2) iron (3) silver (4) zinc

20. Which group contains a liquid that is a nonmetal at STP?
(1) 14 (2) 15 (3) 16 (4) 17

21. Which of these Group 14 elements has the most metallic properties? (1) C (2) Ge (3) Si (4) Sn

22. As the elements in Group 2 are considered in order of increasing atomic number, the atomic radius of each successive element increases. This increase is primarily due to an increase in the number of (1) occupied principal energy levels (2) electrons in the outermost shell (3) neutrons in the nucleus (4) unpaired electrons

23. Which element is classified as a metalloid (semi-metal)?
(1) sulfur (2) silicon (3) barium (4) bromine

24. Which element in Group 1 has the greatest tendency to lose an electron?
(1) cesium (2) rubidium (3) potassium (4) sodium

25. The table below shows some properties of elements A, B, C, and D.

Element	Ionization Energy	Electronegativity	Conductivity of Heat and Electricity
A	low	low	low
B	low	low	high
C	high	high	low
D	high	high	high

Which element is most likely a nonmetal?
(1) A   (2) B   (3) C   (4) D

26. What is the gram formula mass of $Ca_3(PO_4)_2$?
(1) 135 g/mol   (2) 215 g/mol   (3) 278 g/mol   (4) 310. g/mol

27. The gram atomic mass of oxygen is 16.0 grams per mole. How many atoms of oxygen does this mass represent?
(1) 16.0   (2) 32:0   (3) 6.02×10^{23} (4) $2(6.02 \times 10^{23})$

28. Given the *unbalanced* equation: $N_2(g) + H_2(g) \rightarrow NH_3(g)$

When the equation is balanced using smallest whole-number coefficients, the ratio of moles of hydrogen consumed to moles of ammonia produced is   (1) 1:3   (2) 2:3   (3) 3:1   (4) 3:2

29. What is the concentration of a solution of 10. moles of copper (II) nitrate in 5.0 liters of solution?
(1) 0.50 M   (2) 2.0 M   (3) 5.0 M   (4) 10. M

30. Given the balanced equation:
$$Mg(s) + 2HCl(aq) \rightarrow MgCl_2(aq) + H_2(g)$$

At STP, what is the total number of liters of hydrogen gas produced when 3.00 moles of hydrochloric acid solution is completely consumed?
(1) 11.2 L   (2) 22.4 L   (3) 33.6 L   (4) 44.8 L

31. According to Reference Table D, which compound's solubility decreases most rapidly when the temperature increases from 50°C to 70°C?
(1) NH_3 (2) HCl   (3) SO_2 (4) KNO_3

32. Given the reaction at equilibrium:

$$2CO(g) + O_2(g) \rightleftharpoons 2CO_2(g)$$

When the reaction is subjected to stress, a change will occur in the concentration of (1) reactants, only (2) products, only (3) both reactants and products   (4) neither reactants nor products

33. An increase in the temperature of a system at equilibrium favors the (1) endothermic reaction and decreases its rate (2) endothermic reaction and increases its rate (3) exothermic reaction and decreases its rate (4) exothermic reaction and increases its rate

34. Based on Reference Table *L*, which compound, when in aqueous solution, is best conductor of electricity?
    (1) HF (2) H_2S (3) H_2O (4) H_2SO_4

35. A compound that can act as an acid or a base is referred to as
    (1) a neutral substance (2) an amphoteric substance
    (3) a monomer (4) an isomer

Base your answers to questions 36 and 37 on the potential energy diagram of a chemical reaction shown below.

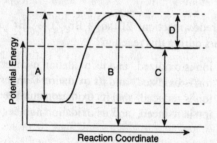

36. Which arrow represents the activation energy for the forward reaction? (1) *A* (2) *B* (3) *C* (4) *D*

37. The forward reaction is best described as an
    (1) exothermic reaction in which energy is released
    (2) exothermic reaction in which energy is absorbed
    (3) endothermic reaction in which energy is released
    (4) endothermic reaction in which energy is absorbed

38. In the reaction $HNO_3 + H_2O \rightleftharpoons H_3O^+ + NO_3^-$, the two Brönsted acids are
    (1) H_2O and HNO_3 (2) H_2O and NO_3^- (3) H_2O and H_3O^+
    (4) HNO_3 and H_3O^+

39. Which substance can be classified as an Arrhenius acid?
    (1) HCl (2) NaCl (3) LiOH (4) KOH

40. How many milliliters of 0.20 M HCl are needed to exactly neutralize 40. milliliters of 0.40 M KOH?
    (1) 20. mL (2) 40. mL (3) 80. mL (4) 160 mL

41. Which 0.1 M solution will turn phenolphthalein pink?
(1) HBr(aq)   (2) CO_2(aq)   (3) LiOH(aq)   (4) CH_3OH(aq)

42. Which compound is an electrolyte?
(1) CH_3OH (2) CH_3COOH (3) $C_3H_5(OH)_3$ (4) $C_{12}H_{22}O_{11}$

43. What is the oxidation number of chlorine in $HClO_4$?
(1) +1   (2) +5   (3) +3   (4) +7

44. Given the redox reaction:

$$Fe^{2+}(aq) + Zn(s) \rightarrow Zn^{2+}(aq) + Fe(s)$$

Which species acts as a reducing agent?
(1) Fe(s)   (2) Fe^{2+}(aq)   (3) Zn(s)   (4) Zn^{2+}(aq)

45. Given the redox reaction: $2I^-(aq) + Br_2(\ell) \rightarrow 2Br^-(aq) + I_2(s)$
What occurs during this reaction?

(1) The I^- ion is oxidized, and its oxidation number increases.
(2) The I^- ion is oxidized, and its oxidation number decreases.
(3) The I^- ion is reduced, and its oxidation number increases.
(4) The I^- ion is reduced, and its oxidation number decreases.

46. Given the reaction: $Zn(s) + 2HCl(aq) \rightarrow ZnCl_2(aq) + H_2(g)$
Which equation represents the correct oxidation half-reaction?

(1) $Zn(s) \rightarrow Zn^{2+} + 2e^-$ (2) $2H^+ + 2e^- \rightarrow H_2(g)$

(3) $Zn^{2+} + 2e \rightarrow Zn(s)$ (4) $2Cl^- \rightarrow Cl_2(g) + 2e^-$

47. According to Reference Table *N*, which redox reaction occurs spontaneously?

(1) $Cu(s) + 2H^+ \rightarrow Cu^{2+} + H_2(g)$
(2) $Mg(s) + 2H^+ \rightarrow Mg^{2+} + H_2(g)$
(3) $2Ag(s) + 2H^+ \rightarrow 2Ag^+ + H_2(g)$
(4) $Hg(\ell) + 2H^+ \rightarrow Hg^{2+} + H_2(g)$

48. Which quantities are conserved in all oxidation-reduction reactions?
(1) charge, only   (2) mass, only   (3) both charge and mass
(4) neither charge nor mass

49. The reaction $CH_2CH_2 + H_2 \rightarrow CH_3CH_3$ is an example of
(1) substitution   (2) addition   (3) esterification   (4) fermentation

50. Given the compound:

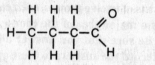

Which structural formula represents an isomer?

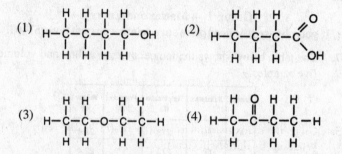

51. In which pair of hydrocarbons does each compound contain only one double bond per molecule?
    (1) C_2H_2 and C_2H_6 (2) C_2H_2 and C_3H_6
    (3) C_4H_8 and C_2H_4 (4) C_6H_6 and C_7H_8

52. Which organic compound is classified as an acid?
    (1) CH_3CH_2COOH (2) CH_3CH_2OH
    (3) $C_{12}H_{22}O_{11}$ (4) $C_6H_{12}O_6$

53. The products of the fermentation of a sugar are ethanol and
    (1) water   (2) oxygen   (3) carbon dioxide   (4) sulfur dioxide

Note that questions 54 through 56 have only three choices.

54. As the temperature of $H_2O(\ell)$ in a closed system decreases, the vapor pressure of the $H_2O(\ell)$
    (1) decreases   (2) increases   (3) remains the same

55. As the number of neutrons in the nucleus of a given atom of an element increases, the atomic number of that element
    (1) decreases   (2) increases   (3) remains the same

56. Given the closed system at equilibrium:

$$CO_2(g) \rightleftharpoons CO_2(aq)$$

As the pressure on the system increases, the solubility of the $CO_2(g)$

(1) decreases   (2) increases   (3) remains the same

Part II

This part consists of twelve groups, each containing five questions. Each group tests a major area of the course. Choose seven of these twelve groups. Be sure that you answer all five questions in each group chosen. Record the answers to these questions on the separate answer sheet in accordance with the directions on the front page of this booklet. [35]

Group 1 — Matter and Energy
If you choose this group, be sure to answer questions 57–61.

57. The table below shows the temperature, pressure, and volume of five samples.

Sample	Substance	Temperature (K)	Pressure (atm)	Volume (L)
A	He	273	1	22.4
B	O_2	273	1	22.4
C	Ne	273	2	22.4
D	N_2	546	2	44.8
E	Ar	546	2	44.8

Which sample contains the same number of molecules as sample A? (1) E   (2) B   (3) C   (4) D

58. The energy absorbed when ammonium chloride dissolves in water can be measured in
(1) degrees   (2) kilocalories   (3) moles per liter   (4) liters per mole

59. At 1 atmosphere of pressure, the steam-water equilibrium occurs at a temperature of
(1) 0 K   (2) 100 K   (3) 273 K   (4) 373 K

60. The graph below represents the uniform cooling of a substance, starting with the substance as a gas above its boiling point.

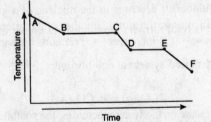

During which interval is the substance completely in the liquid phase?   (1) AB   (2) BC   (3) CD   (4) DE

61. Which two compounds readily sublime at room temperature (25°C)?
    (1) $CO_2(s)$ and $I_2(s)$
    (2) $CO_2(s)$ and $C_6H_{12}O_6(s)$
    (3) $NaCl(s)$ and $I_2(s)$
    (4) $NaCl(s)$ and $C_6H_{12}O_6(s)$

Group 2 — Atomic Structure
If you choose this group, be sure to answer questions 62–66.

62. Which electron configuration is possible for a nitrogen atom in the excited state?
    (1) $1s^2 2s^2 2p^3$ (2) $1s^2 2s^2 2p^3 3s^1$ (3) $1s^2 2s^2 2p^4$ (4) $1s^2 2s^2 2p^2$

63. What is the total amount of energy required to remove the most loosely bound electron from each atom in a mole of gaseous Ca?
    (1) 100 kcal/mol
    (2) 119 kcal/mol
    (3) 141 kcal/mol
    (4) 176 kcal/mol

64. What is the total number of unpaired electrons in an atom of nickel in the ground state?   (1) 0   (2) 2   (3) 3   (4) 4

65. The characteristic bright-line spectrum of an element is produced when its electrons
    (1) form a covalent bond   (2) form an ionic bond   (3) move to a higher energy state   (4) return to a lower energy state

66. Which emanation has *no* mass and *no* charge?
    (1) alpha   (2) beta   (3) gamma   (4) neutron

Group 3 — Bonding
If you choose this group, be sure to answer questions 67–71.

67. Given the incomplete equation:
    $$2N_2O_5(g) \rightarrow$$
    Which set of products completes and balances the incomplete equation?
    (1) $2N_2(g) + 3H_2(g)$ (2) $2N_2(g) + 2O_2(g)$ (3) $4NO_2(g) + O_2(g)$
    (4) $4NO(g) + 5O_2(g)$

68. Which structural formula represents a nonpolar molecule?

    (1)   H — Cl

    (2)   H — O
                  |
                  H

    (3)   H—H

    (4)   H—N—H
                |
                H

69. Compared to the boiling point of H_2S, the boiling point of H_2O is relatively high. Which type of bonding causes this difference?
(1) covalent   (2) hydrogen   (3) ionic   (4) network

70. In which system do molecule-ion attractions exist?
(1) NaCl(aq)   (2) NaCl(s)   (3) $C_6H_{12}O_6$(aq)   (4) $C_6H_{12}O_6$(s)

71. An example of an empirical formula is
(1) C_4H_{10} (2) $C_6H_{12}O_6$ (3) $HC_2H_3O_2$ (4) CH_2O

Group 4—Periodic Table
If you choose this group, be sure to answer questions 72–76.

72. The elements from which two groups of the Periodic Table are most similar in their chemical properties?
(1) 1 and 2   (2) 1 and 17   (3) 2 and 17   (4) 17 and 18

73. Which metal is most likely obtained by the electrolysis of its fused salt?   (1) Au   (2) Ag   (3) Li   (4) Zn

74. Which aqueous solution is colored?
(1) $CuSO_4$(aq)   (2) $BaCl_2$(aq)   (3) KCl(aq)   (4) $MgSO_4$(aq)

75. Because of its high reactivity, which element is *never* found free in nature? (1) O   (2) F   (3) N   (4) Ne

76. Which Group 18 element is most likely to form a compound with the element fluorine? (1) He   (2) Ne   (3) Ar   (4) Kr

Group 5—Mathematics of Chemistry
If you choose this group, be sure to answer questions 77–81.

77. At STP, which gas will diffuse more rapidly than Ne?
(1) He   (2) Ar   (3) Kr   (4) Xe

78. The heat of fusion of a compound is 30.0 calories per gram. What is the total number of calories of heat that must be absorbed by a 15.0-gram sample to change the compound from solid to liquid at its melting point?
(1) 15.0 cal   (2) 45.0 cal   (3) 150. cal   (4) 450. cal

79. Which gas has a density of 1.70 grams per liter at STP?
(1) F_2(g)   (2) He(g)   (3) N_2(g)   (4) SO_2(g)

80. Given the reaction:
$$2C_2H_2(g) + 5O_2(g) \rightarrow 4CO_2(g) + 2H_2O(g)$$

What is the total number of grams of O_2(g) needed to react completely with 0.50 mole of C_2H_2(g)?
(1) 10. g   (2) 40. g   (3) 80. g   (4) 160 g

81. Which statement describes KCl(aq)?
    (1) KCl is the solute in a homogeneous mixture.
    (2) KCl is the solute in a heterogeneous mixture.
    (3) KCl is the solvent in a homogeneous mixture.
    (4) KCl is the solvent in a heterogeneous mixture.

Group 6—Kinetics and Equilibrium

If you choose this group, be sure to answer questions 82–86.

82. According to Reference Table G, which compound will form spontaneously from its elements? (1) ethene (2) hydrogen iodide (3) nitrogen (II) oxide (4) magnesium oxide

83. Given the equilibrium reaction:

$$AgCl(s) \rightleftharpoons Ag^+(aq) + Cl^-(aq)$$

At 25°C, the K_{sp} is equal to
(1) 6.0×10^{-23} (2) 1.8×10^{-10} (3) 1.0×10^{-7} (4) 9.6×10^{-4}

84. Given the reaction:

$$2N_2(g) + O_2(g) \rightleftharpoons 2N_2O(g)$$

Which statement is true when this closed system reaches equilibrium? (1) All of the $N_2(g)$ has been consumed. (2) All of the $O_2(g)$ has been consumed. (3) Pressure changes no longer occur. (4) The forward reaction no longer occurs.

85. Which equation is used to determine the free energy change during a chemical reaction?
    (1) $\Delta G = \Delta H - \Delta S$ (2) $\Delta G = \Delta H + \Delta S$ (3) $\Delta G = \Delta H - T\Delta S$
    (4) $\Delta G = \Delta H + T\Delta S$

86. Which is the correct equilibrium expression for the reaction $2A(g) + 3B(g) \rightleftharpoons C(g) + 3D(g)$?

(1) $K = \dfrac{[2A]+[3B]}{[C]+[3D]}$ (2) $K = \dfrac{[C]+[3D]}{[2A]+[3B]}$

(3) $K = \dfrac{[A]^2[B]^3}{[C][D]^3}$ (4) $K = \dfrac{[C][D]^3}{[A]^2[B]^3}$

Group 7—Acids and Bases

If you choose this group, be sure to answer questions 87–91.

87. According to Reference Table L, what is the conjugate acid of the hydroxide ion (OH^-)?
    (1) O^{2-} (2) H^+ (3) H_2O (4) H_3O^+

88. Which of the following is the *weakest* Brönsted acid?
    (1) NH_4^+ (2) HSO_4^- (3) H_2SO_4 (4) HNO_3

89. Which compound is a salt?
    (1) CH_3OH (2) $C_6H_{12}O_6$ (3) $H_2C_2O_4$ (4) $KC_2H_3O_2$

90. What is the pH of a solution with a hydronium ion concentration of 0.01 mole per liter?
    (1) 1   (2) 2   (3) 10   (4) 14

91. Given the equation: $H^+ + OH^- \rightarrow H_2O$

    Which type of reaction does the equation represent?
    (1) esterification   (2) decomposition   (3) hydrolysis
    (4) neutralization

Group 8—Redox and Electricity

If you choose this group, be sure to answer questions 92–96.

92. Which reduction half-reaction has a standard electrode potential (E^0) of 1.50 volts?

    (1) $Au^{3+} + 3e^- \rightarrow Au(s)$ (2) $Al^{3+} + 3e^- \rightarrow Al(s)$
    (3) $Co^{2+} + 2e^- \rightarrow Co(s)$ (4) $Ca^{2+} + 2e^- \rightarrow Ca(s)$

93. Given the reaction:

    $$2Li(s) + Cl_2(g) \rightarrow 2LiCl(s)$$

    As the reaction takes place, the $Cl_2(g)$ will
    (1) gain electrons   (2) lose electrons
    (3) gain protons   (4) lose protons

94. The diagram below shows the electrolysis of fused KCl.

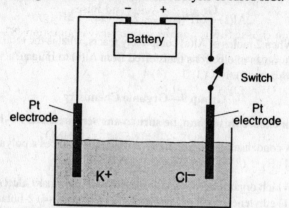

What occurs when the switch is closed? (1) Positive ions migrate toward the anode, where they lose electrons. (2) Positive ions migrate toward the anode, where they gain electrons. (3) Positive ions migrate toward the cathode, where they lose electrons. (4) Positive ions migrate toward the cathode, where they gain electrons.

95. The diagram below represents an electrochemical cell at 298 K and 1 atmosphere.

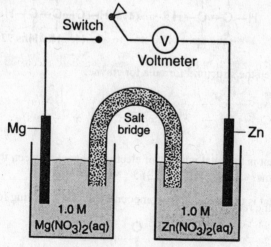

What is the maximum cell voltage (E^0) when the switch is closed?
(1) +1.61 V   (2) –1.61 V   (3) +3.13 V   (4) –3.13 V

96. Given the balanced equation:

$$2Al(s) + 6H^+(aq) \rightarrow 2Al^{3+}(aq) + 3H_2(g)$$

When 2 moles of Al(s) completely reacts, what is the total number of moles of electrons transferred from Al(s) to $H^+(aq)$?
(1) 5   (2) 6   (3) 3   (4) 4

Group 9—Organic Chemistry

If you choose this group, be sure to answer questions 97–101.

97. A condensation polymerization reaction produces a polymer and
(1) H_2 (2) O_2 (3) CO_2 (4) H_2O

98. Which organic compound is classified as a primary alcohol?
(1) ethylene glycol   (2) ethanol   (3) glycerol   (4) 2-butanol

99. Which is the structural formula for 1, 2-ethanediol?

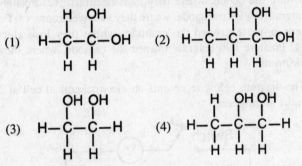

100. Given the structural formula for ethyne:

$$H-C\equiv C-H$$

What is the total number of electrons shared between the carbon atoms?   (1) 6   (2) 2   (3) 3   (4) 4

101. What is the name of the compound with the following formula?

$$H-\overset{\displaystyle H}{\underset{\displaystyle H}{C}}-\overset{\displaystyle O}{C}-\overset{\displaystyle H}{\underset{\displaystyle H}{C}}-H$$

(1) propanone   (2) propanol   (3) propanal   (4) propanoic acid

Group 10—Applications of Chemical Principals

If you choose this group, be sure to answer questions 102–106.

102. Given the overall reaction for the lead-acid battery:

$$Pb + PbO_2 + 2H_2SO_4 \underset{\text{Charge}}{\overset{\text{Discharge}}{\rightleftharpoons}} 2PbSO_4 + 2H_2O$$

Which element changes oxidation state when electric energy is produced?
(1) hydrogen (2) oxygen (3) sulfur (4) lead

103. Which substance is produced by the Haber process?
(1) aluminum (2) ammonia (3) nitric acid (4) sulfuric acid

104. Iron corrodes more easily than aluminum and zinc because aluminum and zinc both (1) are reduced (2) are oxidizing agents (3) form oxides that are self-protective (4) form oxides that are very reactive

105. Which balanced equation represents a cracking reaction?
(1) $2C_3H_6 + 9O_2 \rightarrow 6H_2O + 6CO_2$
(2) $C_{14}H_{30} \rightarrow C_7H_{16} + C_7H_{14}$
(3) $C_{14}H_{28} + Cl_2 \rightarrow C_{14}H_{28}Cl_2$
(4) $C_2H_6 + Cl_2 \rightarrow C_2H_5Cl + HCl$

106. During fractional distillation, hydrocarbons are separated according to their (1) boiling points (2) melting points (3) triple points (4) saturation points

Group 11—Nuclear Chemistry

If you choose this group, be sure to answer questions 107–111.

107. In a fusion reaction, reacting nuclei must collide. Collisions between two nuclei are difficult to achieve because the nuclei are
(1) both negatively charged and repel each other
(2) both positively charged and repel each other
(3) oppositely charged and attract each other
(4) oppositely charged and repel each other

108. A particle accelerator can increase the kinetic energy of
(1) an alpha particle and a beta particle
(2) an alpha particle and a neutron
(3) a gamma ray and a beta particle
(4) a neutron and a gamma ray

109. Which nuclide is a radioisotope used in the study of organic reaction mechanisms?
(1) carbon-12   (2) carbon-14   (3) uranium-235   (4) uranium-238

110. To make nuclear fission more efficient, which device is used in a nuclear reactor to slow the speed of neutrons?   (1) internal shield (2) external shield   (3) control rod   (4) moderator

111. Which equation is an example of artificial transmutation?

(1) $^{238}_{92}U \rightarrow ^{4}_{2}He + ^{234}_{90}Th$ (2) $^{27}_{13}Al + ^{4}_{2}He \rightarrow ^{30}_{15}P + ^{1}_{0}n$

(3) $^{14}_{6}C \rightarrow ^{14}_{7}N + ^{0}_{-1}e$ (4) $^{226}_{88}Ra \rightarrow ^{4}_{2}He + ^{222}_{86}Rn$

Group 12—Laboratory Activities
If you choose this group, be sure to answer questions 112–116.

112. Which measurement contains three significant figures?
(1) 0.08 cm   (2) 0.080 cm   (3) 800 cm   (4) 8.08 cm

113. A student investigated the physical and chemical properties of a sample of an unknown gas and then identified the gas. Which statement represents a conclusion rather than an experimental observation? (1) The gas is colorless. (2) The gas is carbon dioxide. (3) When the gas is bubbled into limewater, the liquid becomes cloudy. (4) When placed in the gas, a flaming splint stops burning.

114. The table below shows properties of four solids, *A*, *B*, *C*, and *D*.

Substance	Melting Point	Conductivity in Solid State	Solubility in Water
A	high	no	soluble
B	high	yes	insoluble
C	high	no	insoluble
D	low	no	insoluble

Which substance could represent diamond, a network solid?
(1) *A*   (2) *B*   (3) *C*   (4) *D*

115. A student obtained the following data to determine the percent by mass of water in a hydrate.

> Mass of empty crucible + cover ....11.70 g
>
> Mass of crucible + cover + hydrated salt before heating.......................14.90 g
>
> Mass of crucible + cover + anhydrous salt after thorough heating ....14.53 g

What is the approximate percent by mass of the water in the hydrated salt?

(1) 2.5%   (2) 12%   (3) 88%   (4) 98%

116. A student wishes to prepare approximately 100 milliliters of an aqueous solution of 6 M HCl using 12 M HCl. Which procedure is correct?

(1) adding 50 mL of 12 M HCl to 50 mL of water while stirring the mixture steadily

(2) adding 50 mL of 12 M HCl to 50 mL of water, and then stirring the mixture steadily

(3) adding 50 mL of water to 50 mL of 12 M HCl while stirring the mixture steadily

(4) adding 50 mL of water to 50 mL of 12 M HCl, and then stirring the mixture steadily

CHEMISTRY
June 18, 1999

Part I
Answer all 56 questions in this part. [65]

Directions (1-56): For *each* statement or question, select the word or expression that, of those given, best completes the statement or answers the question. Record your answer on the separate answer sheet in accordance with the directions on the front page of this booklet.

1. Solid A at 80°C is immersed in liquid B at 60°C. Which statement correctly describes the energy changes between A and B?
   (1) A releases heat and B absorbs heat.
   (2) A absorbs heat and B releases heat.
   (3) Both A and B absorb heat.
   (4) Both A and B release heat.

2. Given the phase equilibrium at a pressure of 1 atmosphere:

$$H_2O(s) \rightleftharpoons H_2O(\ell)$$

   What is the temperature of the equilibrium mixture?
   (1) 273°C   (2) 273 K   (3) 373°C   (4) 373 K

3. Which statement is an identifying characteristic of a mixture?
   (1) A mixture can consist of a single element.
   (2) A mixture can be separated by physical means.
   (3) A mixture must have a definite composition by weight.
   (4) A mixture must be homogeneous.

4. Which phase change is endothermic?
   (1) gas → solid   (3) liquid → solid
   (2) gas → liquid   (4) liquid → gas

5. What volume will a 300.-milliliter sample of a gas at STP occupy when the pressure is doubled at constant temperature?
   (1) 150. mL   (2) 450. mL   (3) 300. mL   (4) 600. mL

6. Each stoppered flask below contains 2 liters of a gas at STP.

2 L
CH_4 (g)

2 L
O_2 (g)

Each gas sample has the same (1) density (2) mass (3) number of molecules (4) number of atoms

7. Which electron configuration represents an atom in the excited state?
(1) $1s^22s^22p^63s^2$ (2) $1s^22s^22p^63s^1$ (3) $1s^22s^22p^6$ (4) $1s^22s^22p^53s^2$

8. Which electron-dot symbol represents an atom of chlorine in the ground state?

(1) $C l \colon$

(3) $\colon \! C l \cdot$

(2) $\cdot C l \cdot$

(4) $\colon C l \colon$

9. What is the total number of electrons in an atom of an element with an atomic number of 18 and a mass number of 40?
(1) 18   (2) 22   (3) 40   (4) 58

10. After bombarding a gold foil sheet with alpha particles, scientists concluded that atoms consist mainly of
(1) electrons (2) empty space (3) protons (4) neutrons

11. Which element has atoms in the ground state with a sublevel that is only half filled?
(1) helium (2) beryllium (3) nitrogen (4) neon

12. Which sublevel contains a total of 5 orbitals?
(1) s (2) p (3) d (4) f

13. In how many days will a 12-gram sample of $^{131}_{53}$I decay, leaving a total of 1.5 grams of the original isotope?
(1) 8.0   (2) 16   (3) 20.   (4) 24

14. Electronegativity is a measure of an atom's ability to
(1) attract the electrons in the bond between the atom and another atom (2) repel the electrons in the bond between the atom and another atom (3) attract the protons of another atom (4) repel the protons of another atom

15. The molecular formula of a compound is represented by X_3Y_6 . What is the empirical formula of this compound?
(1) X_3Y (2) X_2Y (3) XY_2 (4) XY_3

16. Given the unbalanced equation:
$$\_CaSO_4 + \_AlCl_3 \rightarrow \_Al_2(SO_4)_3 + \_CaCl_2$$

What is the coefficient of $Al_2(SO_4)_3$ when the equation is completely balanced using the smallest whole-number coefficients?
(1) 1   (2) 2   (3) 3   (4) 4

17. What is the total number of moles of sulfur atoms in 1 mole of $Fe_2(SO_4)_3$?
(1) 1   (2) 15   (3) 3   (4) 17

18. The elements Li and F combine to form an ionic compound. The electron configurations in this compound are the same as the electron configurations of atoms in Group
(1) 1   (2) 14   (3) 17   (4) 18

19. An element with an electronegativity of 0.9 bonds with an element with an electronegativity of 3.1. Which phrase best describes the bond between these elements? (1) mostly ionic in character and formed between two nonmetals (2) mostly ionic in character and formed between a metal and a nonmetal (3) mostly covalent in character and formed between two nonmetals (4) mostly covalent in character and formed between a metal and a nonmetal

20. The symmetrical structure of the CH_4 molecule is due to the fact that the four single bonds between carbon and hydrogen atoms are directed toward the corners of a
(1) triangle   (2) tetrahedron   (3) square   (4) rectangle

21. As the elements in Group 15 are considered in order of increasing atomic number, which sequence in properties occurs?
(1) nonmetal $\rightarrow$ metalloid $\rightarrow$ metal
(2) metalloid $\rightarrow$ metal $\rightarrow$ nonmetal
(3) metal $\rightarrow$ metalloid $\rightarrow$ nonmetal
(4) metal $\rightarrow$ nonmetal $\rightarrow$ metalloid

22. In which set do the elements exhibit the most similar chemical properties? (1) N, O, and F (2) Hg, Br, and Rn (3) Li, Na, and K (4) Al, Si, and P

23. Which reactant is most likely to have *d* electrons involved in a chemical reaction?
    (1) a halogen   (3) a transition element
    (2) a noble gas   (4) an alkali metal

24. Elements in a given period of the Periodic Table contain the same number of
    (1) protons in the nucleus   (3) electrons in the outermost level
    (2) neutrons in the nucleus   (4) occupied principal energy levels

25. The elements known as the alkaline earth metals are found in Group
    (1) 1   (2) 2   (3) 16   (4) 17

26. The properties of silicon are characteristic of
    (1) a metal, only   (2) a nonmetal, only   (3) both a metal and a nonmetal   (4) neither a metal nor a nonmetal

27. Which substance has the greatest molecular mass?
    (1) H_2O_2 (2) NO   (3) CF_4 (4) I_2

28. According to Reference Table *D*, which of the following substances is *least* soluble in 100 grams of water at 50°C?
    (1) NaCl   (2) KCl   (3) NH_4Cl (4) HCl

29. Which sample contains a total of 6.0×10^{23} atoms?
    (1) 23 g Na   (2) 24 g C   (3) 42 g Kr   (4) 78 g K

30. A 20.-milliliter sample of 0.60 M HCl is diluted with water to a volume of 40. milliliters. What is the new concentration of the solution?
    (1) 0.15 M   (2) 0.60 M   (3) 0.30 M   (4) 1.2 M

31. Given the reaction:
    $$4Al(s) + 3O_2(g) \rightarrow 2Al_2O_3(s)$$

    What is the minimum number of grams of oxygen gas required to produce 1.00 mole of aluminum oxide?
    (1) 32.0 g   (2) 48.0 g   (3) 96.0 g   (4) 192 g

32. The minimum amount of energy required to start a chemical reaction is called
    (1) entropy   (2) enthalpy   (3) free energy   (4) activation energy

33. Beaker *A* contains a 1-gram piece of zinc and beaker *B* contains 1 gram of powdered zinc. If 100 milliliters of 0.1 M HCl is added to each of the beakers, how does the rate of reaction in beaker *A* compare to the rate of reaction in beaker *B*?

(1) The rate in *A* is greater due to the smaller surface area of the zinc. (2) The rate in *A* is greater due to the larger surface area of the zinc. (3) The rate in *B* is greater due to the smaller surface area of the zinc. (4) The rate in *B* is greater due to the larger surface area of the zinc.

34. Based on Reference Table *D*, which amount of a compound dissolved in 100 grams of water at the stated temperature represents a system at equilibrium?

(1) 20 g $KClO_3$ at 80°C   (3) 40 g KCl at 60°C

(2) 40 g KNO_3 at 25°C   (4) 60 g $NaNO_3$ at 40°C

35. Given the reaction at equilibrium:

$$N_2(g) + 3H_2(g) \rightleftharpoons 2NH_3(g) + 22 \text{ kcal}$$

Which stress would cause the equilibrium to shift to the left?

(1) increasing the temperature

(2) increasing the pressure

(3) adding $N_2(g)$ to the system

(4) adding $H_2(g)$ to the system

36. The potential energy diagram of a chemical reaction is shown below.

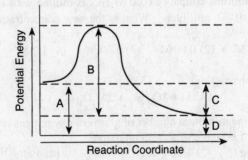

Which arrow represents the part of the reaction most likely to be affected by the addition of a catalyst?

(1) *A*   (2) *B*   (3) *C*   (4) *D*

37. Based on Reference Table *L*, which of the following acids is the strongest electrolyte?
    (1) H_3PO_4 (2) HNO_2 (3) HCl   (2) HF

38. Based on Reference Table *L*, which substance is amphoteric (amphiprotic)?
    (1) HS^- (2) Br^- (3) HBr   (4) H_2S

39. Which type of reaction occurs when 50-milliliter quantities of $Ba(OH)_2(aq)$ and $H_2SO_4(aq)$ are combined?
    (1) hydrolysis (2) ionization (3) hydrogenation (4) neutralization

40. In an acid solution, the $[H^+]$ ion is found to be 1×10^{-2} mole per liter. What is the $[OH^-]$ ion in moles per liter?
    (1) 1×10^{-2} (2) 1×10^{-7} (1) 1×10^{-12} (1) 1×10^{-14}

41. Which statement best describes the solution produced when an Arrhenius acid is dissolved in water?
    (1) The only negative ion in solution is OH^-.
    (2) The only negative ion in solution is HCO_3^-.
    (3) The only positive ion in solution is H^+.
    (4) The only positive ion in solution is NH_4^+.

42. In which reaction is water acting only as a Brönsted-Lowry base?

    (1) $H_2SO_4(aq) + H_2O(\ell) \rightleftharpoons HSO_4^-(aq) + H_3O^+(aq)$

    (2) $NH_3(g) + H_2O(\ell) \rightleftharpoons NH_4^+(aq) + OH^-(aq)$

    (3) $CH_3COO^-(aq) + H_2O(\ell) \rightleftharpoons CH_3COOH(aq) + OH^-(aq)$

    (4) $H_2O(\ell) + H_2O(\ell) \rightleftharpoons H_3O^+(aq) + OH^-(aq)$

43. Which of the following 0.1 M solutions has the *lowest* pH?
    (1) 0.1 M NaOH   (3) 0.1 M NaCl
    (2) 0.1 M CH_3OH (4) 0.1 M HCl

44. The reaction $2H_2O(\ell) \rightarrow 2H_2(g) + O_2(g)$ is forced to occur by use of an externally applied electric current. This procedure is called
    (1) neutralization   (3) electrolysis
    (2) esterification   (4) hydrolysis

45. Given the reaction: $3Sn^{4+}(aq) + 2Cr(s) \rightarrow 3Sn^{2+}(aq) + 2Cr^{3+}(aq)$
    Which half-reaction correctly represents the reduction that occurs?
    (1) $Sn^{4+}(aq) + 2e^- \rightarrow Sn^{2+}(aq)$
    (2) $Sn^{2+}(aq) \rightarrow Sn^{4+}(aq) + 2e^-$
    (3) $Cr(s) \rightarrow Cr^{3+}(aq) + 3e^-$
    (4) $Cr^{3+}(aq) + 3e^- \rightarrow Cr(s)$

46. The oxidation number of nitrogen in N_2 is
    (1) +1   (2) 0   (3) +3   (4) –3

47. Which reaction is an organic reaction?

    (1) $C_3H_8(g) + 5O_2(g) \rightarrow 3CO_2(g) + 4H_2O(g)$
    (2) $2H_2(g) + O_2(g) \rightarrow 2H_2O(g)$
    (3) $3Cu^{2+}(aq) + 2Fe(s) \rightarrow 3Cu(s) + 2Fe^{3+}(aq)$
    (4) $NaOH(aq) + HCl(aq) \rightarrow NaCl(aq) + H_2O(\ell)$

48. Given the reaction:
    $$Zn(s) + 2HCl(aq) \rightarrow ZnCl_2(aq) + H_2(g)$$

    Which substance is oxidized?
    (1) $Zn(s)$ (2) $HCl(aq)$ (3) $Cl^-(aq)$ (4) $H^+(aq)$

49. Which organic compound will dissolve in water to produce a solution that will turn blue litmus red?

    (1)
    ```
       H H
       | |
    H—C—C—H
       | |
       H H
    ```

    (3)
    ```
       H O
       |  ‖
    H—C—C—O—H
       |
       H
    ```

    (2)
    ```
       H     H
       |     |
    H—C—O—C—H
       |     |
       H     H
    ```

    (4)
    ```
       H O H
       |  ‖ |
    H—C—C—C—H
       |    |
       H    H
    ```

50. A redox reaction always demonstrates the conservation of
    (1) mass, only   (3) both mass and charge
    (2) charge, only   (4) neither mass nor charge

51. In which organic reaction is sugar converted to an alcohol and carbon dioxide?
    (1) esterification   (3) substitution
    (2) addition   (4) fermentation

52. Which three compounds belong to the same homologous series?
    (1) CH_4, C_2H_6, C_3H_4 (3) $C_4H_{10}, C_5H_{10}, C_6H_6$
    (2) $C_3H_6, C_4H_8, C_5H_{10}$ (4) C_2H_2, C_3H_4, C_4H_8

53. Which formula represents a saturated compound?
    (1) C_2H_4 (2) C_2H_2 (3) C_3H_6 (4) C_3H_8

    Note that questions 54 through 56 have only three choices.

54. Given the reaction: $2Na(s) + Cl_2(g) \rightarrow 2NaCl(s)$
    As the reactants form products, the stability of the chemical system will
    (1) decrease   (2) increase   (3) remain the same

55. As the elements of Group 1 are considered in order from top to bottom, the first ionization energy of each successive element will
    (1) decrease   (2) increase   (3) remain the same

56. Given the reaction at equilibrium:

    $$4HCl(g) + O_2(g) \rightleftharpoons 2Cl_2(g) + 2H_2O(g)$$

    If the pressure on the system is increased, the concentration of $Cl_2(g)$ will

    (1) decrease   (2) increase   (3) remain the same

Part II

This part consists of twelve groups, each containing five questions. Each group tests a major area of the course. Choose seven of these twelve groups. Be sure that you answer all five questions in each group chosen. Record the answers to these questions on the separate answer sheet in accordance with the directions on the front page of this booklet. [35]

Group 1 — Matter and Energy
If you choose this group, be sure to answer questions 57–61.

57. Which substance can be decomposed by a chemical change?
    (1) ammonia   (2) aluminum   (3) magnesium   (4) manganese

58. The heat required to change 1 gram of a solid at its normal melting point to a liquid at the same temperature is called the heat of
    (1) vaporization   (2) fusion   (3) reaction   (4) formation

59. A real gas would behave most like an ideal gas under conditions of
    (1) low pressure and low temperature
    (2) low pressure and high temperature
    (3) high pressure and low temperature
    (4) high pressure and high temperature

60. The volume of a sample of a gas at 273°C is 200. liters. If the volume is decreased to 100. liters at constant pressure, what will be the new temperature of the gas?
    (1) 0 K   (2) 100. K   (3) 273 K   (4) 546 K

61. The graph below represents the relationship between pressure and volume of a given mass of a gas at constant temperature.

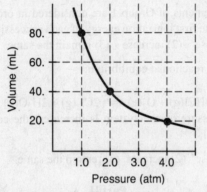

The product of pressure and volume is constant. According to the graph, what is the product in atm•mL?
    (1) 20.   (2) 40.   (3) 60.   (4) 80.

Group 2 — Atomic Structure
If you choose this group, be sure to answer questions 62–66.

62. The mass of a calcium atom is due primarily to the mass of its
    (1) protons, only   (3) protons and neutrons
    (2) neutrons, only   (4) protons and electrons

63. What is the maximum number of electrons that can occupy the fourth principal energy level of an atom?
    (1) 6   (2) 8   (3) 18   (4) 32

64. Which element has no known stable isotope?
    (1) Hg   (2) Po   (3) Se   (4) Zn

65. The characteristic spectral lines of elements are caused when electrons in an excited atom move from
    (1) lower to higher energy levels, releasing energy
    (2) lower to higher energy levels, absorbing energy
    (3) higher to lower energy levels, releasing energy
    (4) higher to lower energy levels, absorbing energy

66. What is the total number of protons contained in the nucleus of a carbon-14 atom?
    (1) 6   (2) 8   (3) 12   (4) 14

Group 3 — Bonding
If you choose this group, be sure to answer questions 67–71.

67. At 298 K, the vapor pressure of H_2O is less than the vapor pressure of CH_3OH because H_2O has
    (1) larger molecules   (3) stronger ionic bonds
    (2) a larger molecular mass   (4) stronger hydrogen bonds

68. The chemical formula $CaCO_3$ is an example of an expression that is
    (1) quantitative, only   (3) both quantitative and qualitative
    (2) qualitative, only   (4) neither quantitative nor qualitative

69. When $NaCl(s)$ is dissolved in $H_2O(\ell)$, the sodium ion is attracted to the water molecule's
    (1) negative end, which is hydrogen
    (2) negative end, which is oxygen
    (3) positive end, which is hydrogen
    (4) positive end, which is oxygen

70. Which electron-dot formula represents a substance that contains a nonpolar covalent bond?

    (1) $[Na]^+ [\overset{x\ x}{\underset{x\ x}{x}Cl x}]^-$ (3) $H \overset{x\ x}{\underset{x\ x}{x}Cl x}$

    (2) $\overset{x\ x}{\underset{x\ x}{x}Cl x} \overset{\bullet\bullet}{\underset{\bullet\bullet}{Cl}}$ (4) $\overset{\bullet\bullet}{\underset{\bullet}{O}x} H$
                                                                                  H

71. What is the correct formula for iron (II) sulfide?
    (1) FeS   (2) $FeSO_3$ (3) Fe_2S_3 (4) $Fe_2(SO_4)_3$

Group 4—Periodic Table
If you choose this group, be sure to answer questions 72–76.

72. Which of the following groups in the Periodic Table contain elements so highly reactive they are never found in the free state?
(1) 1 and 2   (2) 1 and 11   (3) 2 and 15   (4) 11 and 15

73. The presence of which ion usually produces a colored solution?
(1) K^+ (2) F^- (3) Fe^{2+} (4) S^{2-}

74. How does the size of a barium ion compare to the size of a barium atom?
(1) The ion is smaller because it has fewer electrons.
(2) The ion is smaller because it has more electrons.
(3) The ion is larger because it has fewer electrons.
(4) The ion is larger because it has more electrons

75. Which element is brittle in the solid phase and is a *poor* conductor of heat and electricity?
(1) calcium   (2) sulfur   (3) strontium   (4) copper

76. Which halogen can only be prepared by the electrolysis of its fused compounds?   (1) I_2 (2) Cl_2 (3) Br_2 (4) F_2

Group 5—Mathematics of Chemistry
If you choose this group, be sure to answer questions 77–81.

77. What is the mass of 1 mole of a gas that has a density of 2.00 grams per liter at STP?
(1) 11.2 g   (2) 22.4 g   (3) 33.6 g   (4) 44.8 g

78. Dissolving 1 mole of KCl in 1,000 grams of H_2O affects
(1) the boiling point of the H_2O, only
(2) the freezing point of the H_2O, only
(3) both the boiling point and the freezing point of the H_2O
(4) neither the boiling point nor the freezing point of the H_2O

79. The heat of vaporization of a liquid is 320. calories per gram. What is the minimum number of calories needed to change 40.0 grams of the liquid to vapor at the boiling point?
(1) 8.00   (2) 320.   (3) 3,280   (4) 12,800

80. A compound was analyzed and found to contain 75% carbon and 25% hydrogen by mass. What is the compound's empirical formula?   (1) CH (2) CH_2 (3) CH_3 (4) CH_4

81. Which gas diffuses most rapidly at STP?
    (1) Ne   (2) Ar   (3) Cl_2 (4) F_2

Group 6—Kinetics and Equilibrium

If you choose this group, be sure to answer questions 82–86.

82. According to Reference Table G, which compound is spontaneously formed even though the reaction is endothermic?
    (1) ICl(g)   (2) CO_2(g)   (3) $H_2O(\ell)$ (4) Al_2O_3(s)

83. Given the reaction at equilibrium:

    $$BaCrO_4(s) \rightleftharpoons Ba^{2+}(aq) + CrO_4^{2-}(aq)$$

    Which substance, when added to the mixure, will cause an increase in the amount of $BaCrO_4$(s)?
    (1) K_2CO_3 (2) $CaCO_3$ (3) $BaCl_2$ (4) $CaCl_2$

84. At 1 atmosphere and 298 K, which saturated salt solution is most concentrated? [Refer to Reference Table M.]
    (1) $PbCO_3$ (2) $PbCrO_4$ (3) AgBr   (4) AgCl

85. Given the reaction:

    $$A_2B(s) \rightleftharpoons 2A^+(aq) + B^{2-}(aq)$$

    What is the solubility product constant expression (K_{sp}) for this reaction?
    (1) $2[A^+][B^{2-}]$ (3) $[A^+]^2[B^{2-}]$
    (2) $2[A^+] + [B^{2-}]$ (4) $[A^+]^2 + [B^{2-}]$

86. Which factors must be equal when a reversible chemical process reaches equilibrium?
    (1) mass of the reactants and mass of the products
    (2) rate of the forward reaction and rate of the reverse reaction
    (3) concentration of the reactants and concentration of the products
    (4) activation energy of the forward reaction and activation energy of the reverse reaction

Group 7—Acids and Bases

If you choose this group, be sure to answer questions 87–91.

87. The diagram below shows an apparatus used to test the conductivity of various materials.

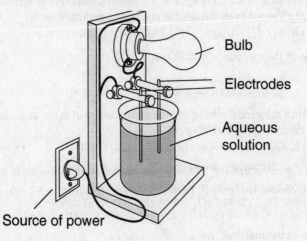

Which aqueous solution will cause the bulb to light?
(1) $C_6H_{12}O_6$(aq)   (3) CH_3OH(aq)
(2) $C_{12}H_{22}O_{11}$(aq)   (4) LiOH(aq)

88. If 50.0 milliliters of 3.0 M HNO_3 completely neutralized 150.0 milliliters of KOH, what was the molarity of the KOH solution?

(1) 1.0 M   (2) 4.5 M   (3) 3.0 M   (4) 6.0 M

89. According to the Arrhenius theory, which list of compounds includes only bases?

(1) KOH, $Ca(OH)_2$, and CH_3OH
(2) KOH, NaOH, and LiOH
(3) LiOH, $Ca(OH)_2$, and $C_2H_4(OH)_2$
(4) NaOH, $Ca(OH)_2$, and CH_3COOH

90. Given the reaction:

$$NH_3(g) + H_2O(\ell) \rightleftharpoons NH_4^+(aq) + OH^-(aq)$$

Which is a conjugate acid-base pair?
(1) $H_2O(\ell)$ and $NH_4^+(aq)$ (3) $NH_3(g)$ and $OH^-(aq)$
(2) $H_2O(\ell)$ and $NH_3(g)$ (4) $NH_3(g)$ and $NH_4^+(aq)$

91. Given the reaction:

$$HC_2H_3O_2(aq) + KOH(aq) \rightarrow KC_2H_3O_2(aq) + H_2O(\ell)$$

The products of this reaction form a salt solution that is
(1) acidic and turns litmus blue
(2) acidic and turns litmus red
(3) basic and turns litmus blue
(4) basic and turns litmus red

Group 8—Redox and Electrochemistry

If you choose this group, be sure to answer questions 92–96.

92. What is the E^0 for the half-reaction $Cu^+ + e^- \rightarrow Cu(s)$?

(1) –0.52 V   (2) –0.34 V   (3) +0.34 V   (4) +0.52 V

93. Given the balanced reaction:
$$2Al(s) + 6H^+(aq) \rightarrow 2Al^{3+}(aq) + 3H_2(g)$$
What is the total number of moles of electrons gained by $H^+(aq)$ when 2 moles of Al(s) is completely reacted?
(1) 6   (2) 2   (3) 3   (4) 12

94. Given the redox reaction:
$$Mg(s) + CuSO_4(aq) \rightarrow MgSO_4(aq) + Cu(s)$$
Which species acts as the oxidizing agent?
(1) Cu(s)   (2) $Cu^{2+}(aq)$ (3) Mg(s)   (4) $Mg^{2+}(aq)$

95. In an electrolytic cell, the negative electrode is called the
(1) anode, at which oxidation occurs
(2) anode, at which reduction occurs
(3) cathode, at which oxidation occurs
(4) cathode, at which reduction occurs

96. The diagram below represents an electro-chemical cell.

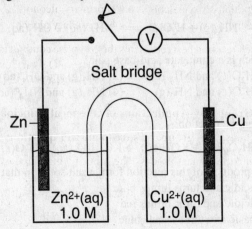

What occurs when the switch is closed?
(1) Zn is reduced.
(2) Cu is oxidized.
(3) Electrons flow from Cu to Zn.
(4) Electrons flow from Zn to Cu.

Group 9—Organic Chemistry

If you choose this group, be sure to answer questions 97–101.

97. Which structural formula represents a primary alcohol?

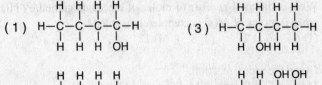

98. Which materials are naturally occurring polymers?
    (1) nylon and cellulose   (3) starch and cellulose
    (2) nylon and polyethylene   (4) starch and polyethylene

99. Which formula represents an isomer of the compound propanoic acid, CH_3CH_2COOH?
    (1) $CH_3CH_2CH_2OH$ (3) $CH_3CH(OH)CH_2OH$
    (2) $CH_3CH_2CH_2COOH$ (4) CH_3COOCH_3

100. The compound 1,2-ethanediol is a
    (1) monohydroxy alcohol (3) primary alcohol
    (2) dihydroxy alcohol (4) secondary alcohol

101. Which reaction best represents the complete combustion of ethene?
    (1) $C_2H_4 + HCl \rightarrow C_2H_5Cl$ (3) $C_2H_4 + 3O_2 \rightarrow 2CO_2 + 2H_2O$
    (2) $C_2H_4 + Cl_2 \rightarrow C_2H_4Cl_2$ (4) $C_2H_4 + H_2O \rightarrow C_2H_5OH$

Group 10—Applications of Chemical Principals

If you choose this group, be sure to answer questions 102–106.

102. During the contact process, the ores of which kind of compounds are burned to produce SO_2?
    (1) bromides (2) carbides (3) phosphides (4) sulfides

103. Given the equation for the overall reaction in a lead-acid storage battery:

$$Pb(s) + PbO_2(s) + 2H_2SO_4(aq) \underset{\text{charge}}{\overset{\text{discharge}}{\rightleftarrows}} 2PbSO_4(s) + 2H_2O(\ell)$$

Which occurs during the charging of the battery?
    (1) The concentration of H_2SO_4 decreases and the number of moles of Pb(s) increases.
    (2) The concentration of H_2SO_4 decreases and the number of moles of $H_2O(\ell)$ increases.
    (3) The concentration of H_2SO_4 increases and the number of moles of Pb(s) decreases.
    (4) The concentration of H_2SO_4 increases and the number of moles of $H_2O(\ell)$ decreases.

104. The corrosion of iron is an example of
    (1) an oxidation-reduction reaction
    (2) an addition reaction
    (3) a substitution reaction
    (4) a neutralization reaction

105. The separation of petroleum into components based on their boiling points is accomplished by
    (1) cracking (3) fractional distillation
    (2) melting (4) addition polymerization

106. Petroleum is a complex mixture of many
    (1) hydrocarbons (3) organic halides
    (2) aldehydes (4) ketones

Group 11—Nuclear Chemistry

If you choose this group, be sure to answer questions 107–111.

107. The diagram below represents a nuclear reaction in which a neutron bombards a heavy nucleus.

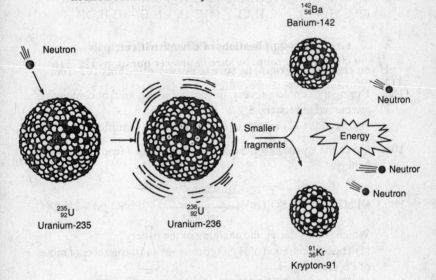

Which type of reaction does the diagram illustrate?
(1) fission   (2) fusion   (3) alpha decay   (4) beta decay

108. Within a nuclear reactor, the purpose of the moderator is to
(1) absorb neutrons in the reactor core
(2) absorb neutrons in the outer containment structure
(3) slow down neutrons in the reactor core
(4) slow down neutrons in the outer containment structure

109. The radioisotope I-131 is used to (1) control nuclear reactors
(2) determine the age of fossils (3) diagnose thyroid disorders
(4) trigger fusion reactors

110. In which list can all particles be accelerated by an electric field?
(1) alpha particles, beta particles, and neutrons
(2) alpha particles, beta particles, and protons
(3) alpha particles, protons, and neutrons
(4) beta particles, protons, and neutrons

111. Given the nuclear reaction: $^{9}_{4}Be + X \rightarrow ^{12}_{6}C + ^{1}_{0}n$

What is the identity of particle X?
(1) alpha particle   (2) beta particle   (3) proton   (4) neutron

Group 12—Laboratory Activities
If you choose this group, be sure to answer questions 112–116.

112. Which set of laboratory equipment would most likely be used with a crucible?

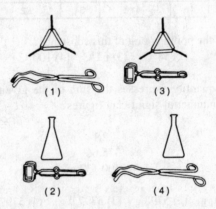

113. A student calculated the percent by mass of water in a sample of $BaCl_2 \cdot 2H_2O$ to be 16.4%, but the accepted value is 14.8%. What was the student's percent error?

(1) $\dfrac{14.8}{16.4} \times 100\%$ (3) $\dfrac{1.6}{14.8} \times 100\%$

(2) $\dfrac{16.4}{14.8} \times 100\%$ (4) $\dfrac{14.8}{1.6} \times 100\%$

114. A student observed the following reaction:
$AlCl_3(aq) + 3NaOH(aq) \rightarrow Al(OH)_3(s) + 3NaCl(aq)$

After the products were filtered, which substance remained on the filter paper?   (1) NaCl   (2) NaOH   (3) $AlCl_3$ (4) $Al(OH)_3$

115. The table below shows the data collected by a student as heat was applied at a constant rate to a solid below its freezing point.

Time (min)	Temperature (°C)	Time (min)	Temperature (°C)
0	20	18	44
2	24	20	47
4	28	22	51
6	32	24	54
8	32	26	54
10	32	28	54
12	35	30	54
14	38	32	58
16	41	34	62

What is the boiling point of this substance?
(1) 32°C   (2) 54°C   (3) 62°C   (4) 100°C

116. Which quantity expresses the sum of the given masses to the correct number of significant figures?

$$22.1 \text{ g}$$
$$375.66 \text{ g}$$
$$+ \, 5400.132 \text{ g}$$

(1) 5800 g   (2) 5798 g   (3) 5797.9 g   (4) 5797.892 g

The University of the State of New York

REGENTS HIGH SCHOOL EXAMINATION

CHEMISTRY

Thursday, June 22, 2000 — 9:15 a.m. to 12:15 p.m., only

The last page of the booklet is the answer sheet. Fold the last page along the perforations and, slowly and carefully, tear off the answer sheet. Then fill in the heading of your answer sheet.

All of your answers are to be recorded on the separate answer sheet. For each question, decide which of the choices given is the best answer. Then on the answer sheet, in the row of numbers for that question, circle with pencil the number of the choice that you have selected. The sample below is an example of the first step in recording your answers.

SAMPLE: ① 2 3 4

If you wish to change an answer, erase your first penciled circle and then circle with pencil the number of the answer you want. After you have completed the examination and you have decided that all of the circled answers represent your best judgment, signal a proctor and turn in all examination material except your answer sheet. Then and only then, place an X in ink in each penciled circle. Be sure to mark only one answer with an X in ink for each question. No credit will be given for any question with two or more X's marked. The sample below indicates how your final choice should be marked with an X in ink.

SAMPLE: ⊗ 2 3 4

The "Reference Tables for Chemistry," which you may need to answer some questions in this examination, are supplied separately. Be certain you have a copy of these reference tables before you begin the examination.

When you have completed the examination, you must sign the statement printed at the end of the answer sheet, indicating that you had no unlawful knowledge of the questions or answers prior to the examination and that you have neither given nor received assistance in answering any of the questions during the examination. Your answer sheet cannot be accepted if you fail to sign this declaration.

DO NOT OPEN THIS EXAMINATION BOOKLET UNTIL THE SIGNAL IS GIVEN.

CHEMISTRY– June 22, 2000 – (2)

Part I

Answer all 56 questions in this part. [65]

Directions (1–56): For *each* statement or question, select the word or expression that, of those given, best completes the statement or answers the question. Record your answer on the separate answer sheet in accordance with the directions on the front page of this booklet.

1 The formula Al_2S_3 represents
(1) an element
(2) a binary compound
(3) a ternary compound
(4) a mixture

2 As the temperature of a gas is increased from 0°C to 10°C at constant pressure, the volume of the gas will
(1) increase by $\frac{1}{273}$ (3) decrease by $\frac{1}{273}$
(2) increase by $\frac{10}{273}$ (4) decrease by $\frac{10}{273}$

3 Water boils at 90°C when the pressure exerted on the liquid is equal to
(1) 50.0 torr (3) 525.8 torr
(2) 100.0 torr (4) 760.0 torr

4 Which species readily sublimes at room temperature?
(1) $CO_2(s)$ (3) $CO_2(g)$
(2) $CO_2(\ell)$ (4) $CO_2(aq)$

5 Which statement best describes all compounds?
(1) They can be decomposed by chemical change.
(2) They can be decomposed by physical means.
(3) They contain at least three elements.
(4) They contain ionic bonds.

6 What is the electron configuration of a Mn atom in the ground state?
(1) $1s^2 2s^2 2p^6 3s^2$
(2) $1s^2 2s^2 2p^6 3s^2 3p^6 3d^5 4s^2$
(3) $1s^2 2s^2 2p^6 3s^2 3p^6 3d^5 4s^1 4p^1$
(4) $1s^2 2s^2 2p^6 3s^2 3p^6 3d^7$

7 Which orbital notation correctly represents a noble gas in the ground state?

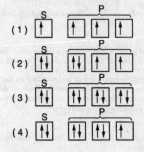

8 Which type of radiation has zero mass and zero charge?
(1) alpha (3) neutron
(2) beta (4) gamma

9 What is the total number of protons and neutrons in an atom of $^{86}_{37}Rb$?
(1) 37 (3) 86
(2) 49 (4) 123

10 What is the total number of valence electrons in an atom of boron in the ground state?
(1) 1 (3) 3
(2) 7 (4) 5

11 What causes the emission of radiant energy that produces characteristic spectral lines?
(1) neutron absorption by the nucleus
(2) gamma ray emission from the nucleus
(3) movement of electrons to higher energy levels
(4) return of electrons to lower energy levels

12 Which atom in the ground state has three half-filled orbitals?

(1) P   (3) Al
(2) Si   (4) Li

13 Which particles may be gained, lost, or shared by an atom when it forms a chemical bond?

(1) protons   (3) neutrons
(2) electrons   (4) nucleons

14 Which molecular formula is correctly paired with its corresponding empirical formula?

(1) CO_2 and CO (3) C_6H_6 and C_2H_2
(2) C_2H_2 and CH_2 (4) P_4O_{10} and P_2O_5

15 Which of the following elements has the strongest attraction for electrons?

(1) boron   (3) oxygen
(2) aluminum   (4) sulfur

16 The table below shows four compounds and the boiling point of each.

Compound	Boiling Point
H_2O	100.°C
H_2S	−60.7°C
H_2Se	−41.5°C
H_2Te	−2.2°C

Which type of molecular attraction accounts for the high boiling point of H_2O?

(1) molecule–ion
(2) ion–ion
(3) hydrogen bonding
(4) van der Waals forces

17 Which elements are both classified as metalloids?

(1) Ge and As   (3) B and C
(2) Bi and Po   (4) Si and P

18 Which electron dot diagram represents a molecule that has a polar covalent bond?

(1) H $\overset{\cdot\cdot}{\underset{\cdot\cdot}{Cl}}$:

(2) Li$^+$ $\left[\overset{\cdot\cdot}{\underset{\cdot\cdot}{Cl}} : \right]^-$

(3) $: \overset{\cdot\cdot}{\underset{\cdot\cdot}{Cl}} \overset{xx}{\underset{xx}{Cl}} x$

(4) K$^+$ $\left[\overset{\cdot\cdot}{\underset{\cdot\cdot}{Cl}} : \right]^-$

19 What is the total number of moles of atoms present in 1 gram formula mass of $Pb(C_2H_3O_2)_2$?

(1) 9   (3) 3
(2) 14   (4) 15

20 Elements in the Periodic Table are arranged according to their

(1) atomic number   (3) relative activity
(2) atomic mass   (4) relative size

21 Which Group 15 element exists as a diatomic molecule at STP?

(1) phosphorus   (3) bismuth
(2) nitrogen   (4) arsenic

22 Which element reacts vigorously with water?

(1) Zn   (3) Fe
(2) Cu   (4) Li

23 Atoms of metals tend to

(1) lose electrons and form negative ions
(2) lose electrons and form positive ions
(3) gain electrons and form negative ions
(4) gain electrons and form positive ions

24 Which halogen is a solid at STP?

(1) Br_2 (3) Cl_2
(2) F_2 (4) I_2

25 What is the total number of molecules in a 0.5-mole sample of He gas?

(1) 6×10^{23} (3) 3×10^{23}
(2) 2×10^{23} (4) 4×10^{23}

26 What occurs as the atomic number of the elements in Period 2 increases?

(1) The nuclear charge of each successive atom decreases, and the covalent radius decreases.
(2) The nuclear charge of each successive atom decreases, and the covalent radius increases.
(3) The nuclear charge of each successive atom increases, and the covalent radius decreases.
(4) The nuclear charge of each successive atom increases, and the covalent radius increases.

27 Given the balanced equation:

$$C_3H_8(g) + 5O_2(g) \rightarrow 3CO_2(g) + 4H_2O(g)$$

What is the total number of liters of $CO_2(g)$ produced when 20.0 liters of $O_2(g)$ are completely consumed?

(1) 12.0 L   (3) 3.00 L
(2) 22.4 L   (4) 5.00 L

28 Given the balanced equation:

$$Fe(s) + CuSO_4(aq) \rightarrow FeSO_4(aq) + Cu(s)$$

What total mass of iron is necessary to produce 1.00 mole of copper?

(1) 26.0 g   (3) 112 g
(2) 55.8 g   (4) 192 g

29 The percent by mass of nitrogen in NH_4NO_3 is closest to

(1) 15%   (3) 35%
(2) 20.%   (4) 60.%

30 What is the molarity of a solution that contains 40. grams of NaOH in 0.50 liter of solution?

(1) 1.0 M   (3) 0.50 M
(2) 2.0 M   (4) 0.25 M

31 Given the system at equilibrium:

$$H_2(g) + F_2(g) \rightleftharpoons 2HF(g) + heat$$

Which change will *not* shift the point of equilibrium?

(1) changing the pressure
(2) changing the temperature
(3) changing the concentration of $H_2(g)$
(4) changing the concentration of $HF(g)$

32 An increase in the surface area of reactants in a heterogeneous reaction will result in

(1) a decrease in the rate of the reaction
(2) an increase in the rate of the reaction
(3) a decrease in the heat of reaction
(4) an increase in the heat of reaction

33 A potential energy diagram of a chemical reaction is shown below.

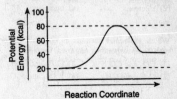

What is the difference between the potential energy of the reactants and the potential energy of the products?

(1) 20. kcal   (3) 60. kcal
(2) 40. kcal   (4) 80. kcal

34 Given the reaction:

$$Zn(s) + 2HCl(aq) \rightarrow$$
$$Zn^{2+}(aq) + 2Cl^-(aq) + H_2(g)$$

If the concentration of the $HCl(aq)$ is increased, the frequency of reacting collisions will

(1) decrease, producing a decrease in the reaction rate
(2) decrease, producing an increase in the reaction rate
(3) increase, producing a decrease in the reaction rate
(4) increase, producing an increase in the reaction rate

35 Two reactant particles collide with proper orientation. The collision will be effective if the particles have

(1) high activation energy
(2) high ionization energy
(3) sufficient kinetic energy
(4) sufficient potential energy

36 Based on Reference Table *D*, what change will cause the solubility of $KNO_3(s)$ to increase?

(1) decreasing the pressure
(2) increasing the pressure
(3) decreasing the temperature
(4) increasing the temperature

37 Which of the following ionization constants (K_a) represents the strongest acid?

(1) $K_a = 1 \times 10^{-14}$ (3) $K_a = 1 \times 10^{-4}$
(2) $K_a = 1 \times 10^{-7}$ (4) $K_a = 1 \times 10^{-2}$

38 Based on Reference Table *L*, which of the following compounds is the *weakest* electrolyte?

(1) HI   (3) H_3PO_4
(2) HNO_3 (4) H_2SO_4

39 Based on Reference Table *L*, which substance can function only as a Brönsted-Lowry acid?

(1) HCl   (3) HCO_3^-
(2) HSO_4^- (4) NH_3

40 If the pH of a solution is 9, the solution is

(1) acidic, which turns phenolphthalein pink
(2) acidic, which turns phenolphthalein colorless
(3) basic, which turns phenolphthalein pink
(4) basic, which turns phenolphthalein colorless

41 Given the reaction:

$$CO_3^{2-} + H_2O \rightleftharpoons HCO_3^- + OH^-$$

Which species is the strongest conjugate base?

(1) CO_3^{2-} (3) HCO_3^-
(2) H_2O (4) OH^-

42 What is the pH of a 0.001 M KOH solution?

(1) 14   (3) 3
(2) 11   (4) 7

43 Which simple oxidation-reduction reaction is *not* correctly balanced?

(1) $Sn(s) + Cu^{2+}(aq) \rightarrow Cu(s) + Sn^{2+}(aq)$
(2) $Ni(s) + Sn^{2+}(aq) \rightarrow Sn(s) + Ni^{2+}(aq)$
(3) $2I^-(aq) + Fe^{3+}(aq) \rightarrow Fe^{2+}(aq) + I_2(s)$
(4) $2I^-(aq) + Hg^{2+}(aq) \rightarrow Hg(\ell) + I_2(s)$

44 In the reaction $H_2S + NH_3 \rightleftharpoons NH_4^+ + HS^-$, the two Brönsted-Lowry bases are

(1) NH_3 and HS^- (3) H_2S and NH_3
(2) NH_3 and NH_4^+ (4) H_2S and HS^-

45 A student wishes to set up an electrochemical cell. The following list of materials and equipment will be used:

• two 250-mL beakers
• wire
• one piece of Zn metal
• 125 mL of 0.10 M $Zn(NO_3)_2$
• voltmeter
• switch
• one piece of Pb metal
• 125 mL of 0.10 M $Pb(NO_3)_2$

For the cell to operate properly, the student will also need

(1) an anode
(2) a cathode
(3) an external path for electrons
(4) a salt bridge

46 Given the cell reaction:

$$Ca(s) + Mg^{2+}(aq) \rightarrow Ca^{2+}(aq) + Mg(s)$$

Which substance is oxidized?

(1) $Ca(s)$ (3) $Ca^{2+}(aq)$
(2) $Mg^{2+}(aq)$ (4) $Mg(s)$

47 Chlorine has an oxidation state of +3 in the compound

(1) HClO   (3) $HClO_3$
(2) $HClO_2$ (4) $HClO_4$

48 Given the cell reaction:

$$Sn(s) + Pb^{2+}(aq) \rightarrow Sn^{2+}(aq) + Pb(s)$$

The reduction half-reaction for this cell is

(1) $Pb^{2+}(aq) + 2e^- \rightarrow Pb(s)$
(2) $Pb(s) \rightarrow Pb^{2+}(aq) + 2e^-$
(3) $Sn^{2+}(aq) + 2e^- \rightarrow Sn(s)$
(4) $Sn(s) \rightarrow Sn^{2+}(aq) + 2e^-$

49 Proteins are produced through the process of

(1) addition   (3) polymerization
(2) substitution   (4) combustion

50 What are the products of a fermentation reaction?

(1) an alcohol and carbon monoxide
(2) an alcohol and carbon dioxide
(3) a salt and water
(4) a salt and an acid

51 Which structural formula represents a saturated hydrocarbon?

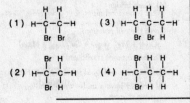

52 Which structural formula represents 1,1-dibromopropane?

53 The principal products of saponification, a reaction between a fat and a base, are soap and

(1) water   (3) carbon dioxide
(2) glycerol   (4) ethyl alcohol

Note that questions 54 through 56 have only three choices.

54 As a solid substance absorbs heat at its melting point, the melting point will

(1) decrease
(2) increase
(3) remain the same

55 Given the redox reaction:

$$2NaCl(\ell) \rightarrow 2Na(\ell) + Cl_2(g)$$

As the Cl^- is oxidized, the oxidation number of chlorine will

(1) decrease
(2) increase
(3) remain the same

56 As energy is released during the formation of a bond, the stability of the chemical system generally will

(1) decrease
(2) increase
(3) remain the same

Part II

This part consists of twelve groups, each containing five questions. Each group tests a major area of the course. Choose seven of these twelve groups. Be sure that you answer all five questions in each group chosen. Record the answers to these questions on the separate answer sheet in accordance with the directions on the front page of this booklet. [35]

Group 1 — Matter and Energy

If you choose this group, be sure to answer questions 57–61.

57 Which term represents a form of energy?

(1) heat   (3) kilocalorie
(2) degree   (4) temperature

58 Which change of phase is exothermic?

(1) solid to liquid   (3) solid to gas
(2) gas to liquid   (4) liquid to gas

59 A gas sample consisting of 2 moles of hydrogen and 1 mole of oxygen is collected over water at 29°C and 750 torr. What is the partial pressure of the hydrogen in the sample?

(1) 240 torr   (3) 720 torr
(2) 480 torr   (4) 750 torr

60 What is the equilibrium temperature of an ice-water mixture at a pressure of 1 atmosphere?

(1) 0°C   (3) 100°C
(2) 32°C   (4) 273°C

61 The list below shows four samples: A, B, C, and D.

(A) HCl(aq)
(B) NaCl(aq)
(C) HCl(g)
(D) NaCl(s)

Which samples are substances?

(1) A and B   (3) C and B
(2) A and C   (4) C and D

Group 2 — Atomic Structure

If you choose this group, be sure to answer questions 62–66.

62 Which subatomic particles have a mass of approximately 1 atomic mass unit each?

(1) proton and electron
(2) proton and neutron
(3) neutron and positron
(4) electron and positron

63 Which atoms are isotopes of the same element?

(1) $^{24}_{12}X$ and $^{25}_{12}X$ (3) $^{31}_{15}X$ and $^{32}_{16}X$
(2) $^{20}_{10}X$ and $^{20}_{11}X$ (4) $^{31}_{19}X$ and $^{31}_{19}X$

64 What is the total number of grams of a 32-gram sample of ^{32}P remaining after 71.5 days of decay?

(1) 1.0 g   (3) 8.0 g
(2) 2.0 g   (4) 4.0 g

65 Experiments with gold foil indicated that atoms

(1) usually have a uniform distribution of positive charges
(2) usually have a uniform distribution of negative charges
(3) contain a positively charged, dense center
(4) contain a negatively charged, dense center

66 What is the total number of completely filled sublevels found in an atom of krypton in the ground state?

(1) 10   (3) 8
(2) 2   (4) 4

Group 3 — Bonding

If you choose this group, be sure to answer questions 67–71.

67 Which atom will form an ionic bond with a Br atom?

(1) N   (3) O
(2) Li   (4) C

68 Given the unbalanced equation:

$$\_\_\_N_2(g) + \_\_\_O_2(g) \rightarrow \_\_\_N_2O_5(g)$$

When the equation is balanced using *smallest* whole numbers, the coefficient of $N_2(g)$ will be

(1) 1   (3) 5
(2) 2   (4) 4

69 Which molecule is polar and contains polar bonds?

(1) CCl_4 (3) N_2
(2) CO_2 (4) NH_3

70 The *strongest* van der Waals forces of attraction exist between molecules of

(1) I_2 (3) Cl_2
(2) Br_2 (4) F_2

71 Which diagram best illustrates the ion-molecule attractions that occur when the ions of NaCl(s) are added to water?

(1)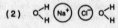

(2)

(3)

(4)

Group 4 — Periodic Table

If you choose this group, be sure to answer questions 72–76.

72 An element has a first ionization energy of 314 kilocalories/mole and an electronegativity of 3.5. It is classified as a

(1) metal   (3) metalloid
(2) nonmetal   (4) halogen

73 At which location in the Periodic Table would the most active metallic element be found?

(1) in Group 1 at the top
(2) in Group 1 at the bottom
(3) in Group 17 at the top
(4) in Group 17 at the bottom

74 Which set of properties is most characteristic of transition elements?

(1) colorless ions in solution, multiple positive oxidation states
(2) colorless ions in solution, multiple negative oxidation states
(3) colored ions in solution, multiple positive oxidation states
(4) colored ions in solution, multiple negative oxidation states

75 An atom with the electron configuration $1s^2 2s^2 2p^6 3s^2$ would most likely

(1) decrease in size as it forms a positive ion
(2) increase in size as it forms a positive ion
(3) decrease in size as it forms a negative ion
(4) increase in size as it forms a negative ion

76 The properties of carbon are expected to be most similar to those of

(1) boron   (3) silicon
(2) aluminum   (4) phosphorus

Group 5 — Mathematics of Chemistry

If you choose this group, be sure to answer questions 77–81.

77 The stoppered tubes below, labeled A through D, each contain a different gas.

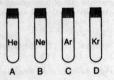

A   B   C   D

When the tubes are unstoppered at the same time and under the same conditions of temperature and pressure, from which tube will gas diffuse at the fastest rate?

(1) A   (3) C
(2) B   (4) D

78 A compound whose empirical formula is NO_2 could have a molecular mass of

(1) 23   (3) 92
(2) 39   (4) 120

79 The density of a gas is 1.43 grams per liter at STP. The mass of 1 mole of this gas is equal to

(1) 1.43 g   (3) 22.4 g
(2) 15.7 g   (4) 32.0 g

80 What is the total number of kilocalories required to boil 100. grams of water at 100°C and 1 atmosphere? [Refer to Reference Table A.]

(1) 1.80 kcal   (3) 53.9 kcal
(2) 18.0 kcal   (4) 539 kcal

81 Which property of a distilled water solution will *not* be affected by adding 50 mL of $CH_3OH(\ell)$ to 100 mL of the water solution at 25°C?

(1) conductivity   (3) freezing point
(2) vapor pressure   (4) boiling point

Group 6 — Kinetics and Equilibrium

If you choose this group, be sure to answer questions 82–86.

82 Based on Reference Table M, which compound has a K_{sp} closest to the K_{sp} of $PbCrO_4$?

(1) Ag_2CrO_4
(2) $AgBr$
(3) $ZnCO_3$
(4) $PbCl_2$

83 Adding a catalyst to a chemical reaction changes the rate of reaction by causing

(1) a decrease in the activation energy
(2) an increase in the activation energy
(3) a decrease in the heat of reaction
(4) an increase in the heat of reaction

84 Which change in a sample of water is accompanied by the greatest increase in entropy?

(1) $H_2O(\ell)$ at 100°C is changed to $H_2O(g)$ at 200°C.
(2) $H_2O(g)$ at 100°C is changed to $H_2O(g)$ at 200°C.
(3) $H_2O(s)$ at –100°C is changed to $H_2O(s)$ at 0°C.
(4) $H_2O(s)$ at –100°C is changed to $H_2O(\ell)$ at 0°C.

85 According to Reference Table G, which reaction spontaneously forms a compound from its elements?

(1) $H_2(g) + I_2(g) \rightarrow 2HI(g)$
(2) $2H_2(g) + O_2(g) \rightarrow 2H_2O(g)$
(3) $N_2(g) + O_2(g) \rightarrow 2NO(g)$
(4) $N_2(g) + 2O_2(g) \rightarrow 2NO_2(g)$

86 Given the solution at equilibrium:

$$CaSO_4(s) \rightleftharpoons Ca^{2+}(aq) + SO_4^{2-}(aq)$$

When Na_2SO_4 is added to the system, how will the equilibrium shift?

(1) The amount of $CaSO_4(s)$ will decrease, and the concentration of $Ca^{2+}(aq)$ will decrease.
(2) The amount of $CaSO_4(s)$ will decrease, and the concentration of $Ca^{2+}(aq)$ will increase.
(3) The amount of $CaSO_4(s)$ will increase, and the concentration of $Ca^{2+}(aq)$ will decrease.
(4) The amount of $CaSO_4(s)$ will increase, and the concentration of $Ca^{2+}(aq)$ will increase.

Group 7 — Acids and Bases

If you choose this group, be sure to answer questions 87–91.

87 When HCl is dissolved in water, the only positive ion present in the solution is the

(1) hydrogen ion   (3) hydride ion
(2) hydroxide ion   (4) chloride ion

88 Given the reaction:

$$2NaOH + H_2SO_4 \rightarrow Na_2SO_4 + 2H_2O$$

How many milliliters of 1 M NaOH are needed to exactly neutralize 100 milliliters of 1 M H_2SO_4?

(1) 50 mL   (3) 300 mL
(2) 200 mL   (4) 400 mL

89 According to the Brönsted-Lowry theory, H_2O is considered to be a base when it

(1) donates an electron   (3) donates a proton
(2) accepts an electron   (4) accepts a proton

90 Given the neutralization reaction:

$$H_2SO_4 + 2KOH \rightarrow K_2SO_4 + 2HOH$$

Which compound is a salt?

(1) KOH   (3) K_2SO_4
(2) H_2SO_4 (4) HOH

91 Which acid-base pair will always undergo a reaction that produces a neutral solution?

(1) a weak acid and a weak base
(2) a weak acid and a strong base
(3) a strong acid and a weak base
(4) a strong acid and a strong base

Group 8 — Redox and Electrochemistry

If you choose this group, be sure to answer questions 92–96.

Base your answers to questions 92 and 93 on the diagram of a chemical cell and the equation shown below. The reaction occurs at 1 atmosphere and 298 K.

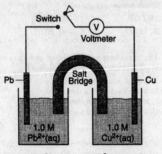

$Pb(s) + Cu^{2+}(aq) \longrightarrow Pb^{2+}(aq) + Cu(s)$

92 When the switch is closed, the cell voltage (E^0) is

(1) –0.21 V   (3) –0.47 V
(2) +0.21 V   (4) +0.47 V

93 Which change occurs when the switch is closed?

(1) Pb is oxidized, and electrons flow to the Cu electrode.
(2) Pb is reduced, and electrons flow to the Cu electrode.
(3) Cu is oxidized, and electrons flow to the Pb electrode.
(4) Cu is reduced, and electrons flow to the Pb electrode.

94 Based on Reference Table *N*, the standard electrode potential for the reduction of gold (III) ions is

(1) +1.50 V   (3) –0.80 V
(2) +0.80 V   (4) –1.50 V

95 In an electrolytic cell, a negative ion will migrate to and undergo oxidation at the

(1) anode, which is negatively charged
(2) anode, which is positively charged
(3) cathode, which is negatively charged
(4) cathode, which is positively charged

96 Given the unbalanced equation:

$$\_\_ Ag_2S + 8HNO_3 \rightarrow$$
$$\_\_ AgNO_3 + 2NO + \_\_ S + \_\_ H_2O$$

What is the coefficient of Ag_2S when the equation is completely balanced using the *smallest* whole numbers?

(1) 6   (3) 3
(2) 2   (4) 4

Group 9 — Organic Chemistry

If you choose this group, be sure to answer questions 97–101.

97 Given the compound:

$$H - C = C - H$$

with H, H above the carbons.

The symbol $=$ represents

(1) one pair of shared electrons
(2) two pairs of shared electrons
(3) a single covalent bond
(4) a coordinate covalent bond

98 An organic compound containing one or more OH groups as the only functional group is classified as an

(1) aldehyde   (3) ester
(2) alcohol   (4) ether

99 The reaction during which monomers are combined and water is released is called

(1) saponification
(2) neutralization
(3) addition polymerization
(4) condensation polymerization

100 One molecule of glycerol contains a total of

(1) two –OH groups
(2) two –CH_3 groups
(3) three –OH groups
(4) three –CH_3 groups

101 What is the general formula for an ether?

(1) R—OH   (3) R—O—R′

(2) R—C—R′ with O double bond below   (4) R—C with O and H

Group 10 — Applications of Chemical Principles

If you choose this group, be sure to answer questions 102–106.

102 Which type of chemical reaction occurs in a lead-acid battery?

(1) addition   (3) esterification
(2) substitution   (4) oxidation-reduction

103 Which metals are obtained by electrolysis of their fused salts?

(1) K and Ca   (3) Cu and Zn
(2) K and Cr   (4) Cu and Hg

104 Given the redox reaction:

$$2NiOOH + Cd \underset{charge}{\overset{discharge}{\rightleftharpoons}} 2Ni(OH)_2 + Cd(OH)_2$$

Which species is oxidized during discharge?

(1) Cd   (3) $Ni(OH)_3$
(2) Cd^{2+} (4) $Ni(OH)_2$

105 Petroleum is primarily a mixture of

(1) alcohol molecules
(2) ester molecules
(3) hydrocarbon molecules
(4) organic acid molecules

106 By which process is petroleum separated into its components according to their different boiling points?

(1) contact process
(2) Haber process
(3) fractional distillation
(4) cracking

Group 11 — Nuclear Chemistry

If you choose this group, be sure to answer questions 107–111.

107 Organic molecules react to form a product. These reactions may be studied by using

(1) Sr-90   (3) N-16
(2) Co-60   (4) C-14

108 In a fission reactor, the speed of the neutrons may be decreased by

(1) a moderator   (3) a fuel rod
(2) an accelerator   (4) shielding

109 Which statement explains why fusion reactions are difficult to initiate?

(1) Positive nuclei attract each other.
(2) Positive nuclei repel each other.
(3) Neutrons prevent nuclei from getting close enough to fuse.
(4) Electrons prevent nuclei from getting close enough to fuse.

110 A particle accelerator is used to provide charged particles with sufficient

(1) kinetic energy to penetrate a nucleus
(2) kinetic energy to penetrate an electron cloud
(3) potential energy to penetrate a nucleus
(4) potential energy to penetrate an electron cloud

111 In which reaction is mass converted to energy by the process of fission?

(1) $^{14}_{7}N + ^{1}_{0}n \rightarrow ^{14}_{6}C + ^{1}_{1}H$

(2) $^{235}_{92}U + ^{1}_{0}n \rightarrow ^{87}_{35}Br + ^{146}_{57}La + 3^{1}_{0}n$

(3) $^{226}_{88}Ra \rightarrow ^{222}_{86}Rn + ^{4}_{2}He$

(4) $^{2}_{1}H + ^{2}_{1}H \rightarrow ^{4}_{2}He$

Group 12 — Laboratory Activities

If you choose this group, be sure to answer questions 112–116.

112 Which piece of glassware is used for accurately measuring volumes of an acid and a base during a titration?

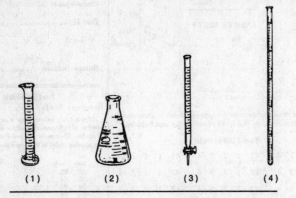

(1)   (2)   (3)   (4)

113 The results of testing a colorless solution with three indicators are shown in the table below.

Indicator	Result
red litmus	blue
blue litmus	blue
phenolphthalein	pink

Which formula could represent the solution tested?

(1) NaOH(aq)

(2) HCl(aq)

(3) $C_6H_{12}O_6$(aq)

(4) $C_{12}H_{22}O_{11}$(aq)

114 What is the product of (2.324 cm × 1.11 cm) expressed to the correct number of significant figures?

(1) 2.58 cm^2

(2) 2.5780 cm^2

(3) 2.5796 cm^2

(4) 2.57964 cm^2

115 A student determined the percentage of water of hydration in $BaCl_2 \cdot 2H_2O$ by using the data in the table below.

Quantity Measured	Value Obtained
mass of $BaCl_2 \cdot 2H_2O$	3.80 grams
mass of $BaCl_2$	3.20 grams
% of water calculated	15.79%

The accepted percentage value for the water of hydration is 14.75%. What is the student's percent error?

(1) 1.04%

(2) 6.00%

(3) 6.59%

(4) 7.05%

116 By which process is a precipitate most easily separated from the liquid in which it is suspended?

(1) neutralization

(2) distillation

(3) condensation

(4) filtration

The University of the State of New York

REGENTS HIGH SCHOOL EXAMINATION

CHEMISTRY

Thursday, June 22, 2000 — 9:15 a.m. to 12:15 p.m., only

ANSWER SHEET

☐ Male

Student .... Sex: ☐ Female

Teacher ....

School ....

Record all of your answers on this answer sheet in accordance with the instructions on the front cover of the test booklet.

Part I (65 credits)

1	1	2	3	4	21	1	2	3	4	41	1	2	3	4
2	1	2	3	4	22	1	2	3	4	42	1	2	3	4
3	1	2	3	4	23	1	2	3	4	43	1	2	3	4
4	1	2	3	4	24	1	2	3	4	44	1	2	3	4
5	1	2	3	4	25	1	2	3	4	45	1	2	3	4
6	1	2	3	4	26	1	2	3	4	46	1	2	3	4
7	1	2	3	4	27	1	2	3	4	47	1	2	3	4
8	1	2	3	4	28	1	2	3	4	48	1	2	3	4
9	1	2	3	4	29	1	2	3	4	49	1	2	3	4
10	1	2	3	4	30	1	2	3	4	50	1	2	3	4
11	1	2	3	4	31	1	2	3	4	51	1	2	3	4
12	1	2	3	4	32	1	2	3	4	52	1	2	3	4
13	1	2	3	4	33	1	2	3	4	53	1	2	3	4
14	1	2	3	4	34	1	2	3	4	54	1	2	3
15	1	2	3	4	35	1	2	3	4	55	1	2	3
16	1	2	3	4	36	1	2	3	4	56	1	2	3
17	1	2	3	4	37	1	2	3	4
18	1	2	3	4	38	1	2	3	4
19	1	2	3	4	39	1	2	3	4
20	1	2	3	4	40	1	2	3	4

Your answers for Part II should be placed in the proper spaces on the back of this sheet.

FOR TEACHER USE ONLY

Credits

Part I ....
(Use table below)

Part II ....

Total ....

Rater's Initials: ....

Part I Credits

Directions to Teacher:

In the table below, draw a circle around the number of right answers and the adjacent number of credits. Then write the number of credits (not the number right) in the space provided above.

No. Right	Credits	No. Right	Credits
56	65	28	41
55	64	27	40
54	63	26	39
53	62	25	39
52	62	24	38
51	61	23	37
50	60	22	36
49	59	21	35
48	58	20	34
47	57	19	33
46	56	18	33
45	56	17	32
44	55	16	31
43	54	15	30
42	53	14	29
41	52	13	27
40	51	12	25
39	51	11	23
38	50	10	21
37	49	9	19
36	48	8	17
35	47	7	14
34	46	6	12
33	45	5	10
32	45	4	8
31	44	3	6
30	43	2	4
29	42	1	2
		0	0

No. right ....

Part II (35 credits)

Answer the questions in only seven of the twelve groups in this part. Be sure to mark the answers to the groups of questions you choose in accordance with the instructions on the front cover of the test booklet. Leave blank the five groups of questions you do not choose to answer.

Group 1 Matter and Energy	Group 2 Atomic Structure	Group 3 Bonding	Group 4 Periodic Table
57 1 2 3 4	62 1 2 3 4	67 1 2 3 4	72 1 2 3 4
58 1 2 3 4	63 1 2 3 4	68 1 2 3 4	73 1 2 3 4
59 1 2 3 4	64 1 2 3 4	69 1 2 3 4	74 1 2 3 4
60 1 2 3 4	65 1 2 3 4	70 1 2 3 4	75 1 2 3 4
61 1 2 3 4	66 1 2 3 4	71 1 2 3 4	76 1 2 3 4

Group 5 Mathematics of Chemistry	Group 6 Kinetics and Equilibrium	Group 7 Acids and Bases	Group 8 Redox and Electrochemistry
77 1 2 3 4	82 1 2 3 4	87 1 2 3 4	92 1 2 3 4
78 1 2 3 4	83 1 2 3 4	88 1 2 3 4	93 1 2 3 4
79 1 2 3 4	84 1 2 3 4	89 1 2 3 4	94 1 2 3 4
80 1 2 3 4	85 1 2 3 4	90 1 2 3 4	95 1 2 3 4
81 1 2 3 4	86 1 2 3 4	91 1 2 3 4	96 1 2 3 4

Group 9 Organic Chemistry	Group 10 Applications of Chemical Principles	Group 11 Nuclear Chemistry	Group 12 Laboratory Activities
97 1 2 3 4	102 1 2 3 4	107 1 2 3 4	112 1 2 3 4
98 1 2 3 4	103 1 2 3 4	108 1 2 3 4	113 1 2 3 4
99 1 2 3 4	104 1 2 3 4	109 1 2 3 4	114 1 2 3 4
100 1 2 3 4	105 1 2 3 4	110 1 2 3 4	115 1 2 3 4
101 1 2 3 4	106 1 2 3 4	111 1 2 3 4	116 1 2 3 4

I do hereby affirm, at the close of this examination, that I had no unlawful knowledge of the questions or answers prior to the examination and that I have neither given nor received assistance in answering any of the questions during the examination.

_____

Signature